Discrete Distributions in Engineering and the Applied Sciences

Synthesis Lectures on Mathematics and Statistics

Editor
Steven G. Krantz, *Washington University, St. Louis*

Analytical Techniques for Solving Nonlinear Partial Differential Equations
Daniel J. Arrigo
2019

Aspects of Differential Geometry IV
Esteban Calviño-Louzao, Eduardo García-Río, Peter Gilkey, JeongHyeong Park, and Ramón Vázquez-Lorenzo
2019

Symmetry Problems. Thne Navier–Stokes Problem.
Alexander G. Ramm
2019
An Introduction to Partial Differential Equations
Daniel J. Arrigo
2017

Numerical Integration of Space Fractional Partial Differential Equations: Vol 2 – Applicatons from Classical Integer PDEs
Younes Salehi and William E. Schiesser
2017

Numerical Integration of Space Fractional Partial Differential Equations: Vol 1 – Introduction to Algorithms and Computer Coding in R
Younes Salehi and William E. Schiesser
2017

Aspects of Differential Geometry III
Esteban Calviño-Louzao, Eduardo García-Río, Peter Gilkey, JeongHyeong Park, and Ramón Vázquez-Lorenzo
2017

The Fundamentals of Analysis for Talented Freshmen
Peter M. Luthy, Guido L. Weiss, and Steven S. Xiao
2016

Aspects of Differential Geometry II
Peter Gilkey, JeongHyeong Park, Ramón Vázquez-Lorenzo
2015

Aspects of Differential Geometry I
Peter Gilkey, JeongHyeong Park, Ramón Vázquez-Lorenzo
2015

Statistics is Easy!
Dennis Shasha and Manda Wilson
2008

A Gyrovector Space Approach to Hyperbolic Geometry
Abraham Albert Ungar
2008

Discrete Distributions in Engineering and the Applied Sciences
Rajan Chattamvelli and Ramalingam Shanmugam

ISBN: 978-3-031-01297-6   paperback
ISBN: 978-3-031-02425-2   ebook
ISBN: 978-3-031-00271-7   hardcover

DOI 10.1007/978-3-031-02425-2

A Publication in the Springer series
SYNTHESIS LECTURES ON MATHEMATICS AND STATISTICS

Lecture #34
Series Editor: Steven G. Krantz, *Washington University, St. Louis*
Series ISSN
Print 1938-1743   Electronic 1938-1751

Discrete Distributions in Engineering and the Applied Sciences

Rajan Chattamvelli
VIT University, Vellore, Tamil Nadu

Ramalingam Shanmugam
Texas State University, San Marcos, Texas

SYNTHESIS LECTURES ON MATHEMATICS AND STATISTICS #34

ABSTRACT

This is an introductory book on discrete statistical distributions and its applications. It discusses only those that are widely used in the applications of probability and statistics in everyday life. The purpose is to give a self-contained introduction to classical discrete distributions in statistics. Instead of compiling the important formulas (which are available in many other textbooks), we focus on important applications of each distribution in various applied fields like bioinformatics, genomics, ecology, electronics, epidemiology, management, reliability, etc., making this book an indispensable resource for researchers and practitioners in several scientific fields. Examples are drawn from different fields. An up-to-date reference appears at the end of the book.

Chapter 1 introduces the basic concepts on random variables, and gives a simple method to find the mean deviation (MD) of discrete distributions. The Bernoulli and binomial distributions are discussed in detail in Chapter 2. A short chapter on discrete uniform distribution appears next. The next two chapters are on geometric and negative binomial distributions. Chapter 6 discusses the Poisson distribution in-depth, including applications in various fields. Chapter 7 is on hypergeometric distribution. As most textbooks in the market either do not discuss, or contain only brief description of the negative hypergeometric distribution, we have included an entire chapter on it. A short chapter on logarithmic series distribution follows it, in which a theorem to find the kth moment of logarithmic distribution using $(k-1)$th moment of zero-truncated geometric distribution is presented. The last chapter is on multinomial distribution and its applications.

The primary users of this book are professionals and practitioners in various fields of engineering and the applied sciences. It will also be of use to graduate students in statistics, research scholars in science disciplines, and teachers of statistics, biostatistics, biotechnology, education, and psychology.

Any suggestions for improvement are welcome. All comments can be sent to `rajan.chattamvelli@vit.ac.in`, and will be incorporated promptly in subsequent editions.

KEYWORDS

additivity, algorithms, biostatistics, biotechnology, capture-recapture model, combinatorics, electronics, generating functions, highway engineering, mean deviation, negative hypergeometric distribution, Poisson limit theorem, probability distributions, quality control, random variables, special functions, Stirling distributions, Stirling numbers, survival function, truncated distribution, variance generating function, zero-inflated distributions

Contents

List of Figures

List of Tables

Glossary

Term	Meaning
BBD	Beta-binomial distribution
CDFGF	Cumulative Distribution Function GF
CGF	Cumulant Generating Function
ChF	Characteristic Function
CMGF	Central Moment Generating Function
CMR	Capture–Mark–Recapture model
CV	Coefficient of Variation
DUNI	Discrete Uniform Distribution
EDI	Excess Dispersion Index
EGF	Exponential Generating Function
FCGF	Factorial Cumulant Generating Function
FMGF	Factorial Moment Generating Function
FPCF	Finite Population Correction Factor
GF	Generating Function
HGD	Hypergeometric Distribution
IID	Independently and Identically distributed
IJR	Incidence Jump Rate
LSD	Logarithmic Series Distribution
MDGF	Mean Deviation Generating Function
MGF	Moment Generating Function
NBINO	Negative Binomial Distribution
NHGD	Negative Hypergeometric Distribution
PGF	Probability Generating Function
PMF	Probability Mass Function
SF	Survival Function
SFGF	Survival Function Generating Function
SNFK	Stirling Number of First Kind
SNSK	Stirling Number of Second Kind
VGF	Variance Generating Function
ZINB	Zero-Inflated Negative Binomial
ZTH	Zero-Truncated Hypergeometric Distribution

CHAPTER 1

Discrete Random Variables

After finishing the chapter, readers will be able to . . .

- Understand binomial theorem and its forms

- Comprehend basic properties of random variables

- Apply the Power method to find mean deviation of discrete distributions

- Explore truncated and inflated distributions

1.1 INTRODUCTION

This book discusses popular discrete distributions and its applications in engineering and the applied sciences. The literature on these distributions are extensive and ever-increasing. Hence, only the most important results that are of practical interest to engineers and researchers in various fields are included. Although the basic properties of these distributions are already available in many excellent monographs (Balakrishnan and Nevzorov (2003) [5]; Forbes, Evans, Hasting, and Peacock (2011) [26]; Johnson, Kemp, and Kotz, (2004) [39]) and online encyclopedias (like wikipedia, scholarpedia, Wolfram, etc.), these are seldom sufficient to apply them to practical problems. Moreover, engineers and professionals working in some fields usually encounter only a few of the distributions for which probability mass function (PMF), cumulative distribution function (CDF), or survival function (SF) (complement of CDF) are needed for special variable values.

1.1.1 DISCRETE MODELS

A model is a simplified description of a real-life phenomena. It may denote a system that simulates a process or activity in engineering sciences. A model is called deterministic if it produces a unique output for a given input. Otherwise it is called a stochastic model. The input as well as output of stochastic models can be random. There are hundreds of models available for a designer—biological models, mathematical models, statistical models, stochastic models, and machine-learning models, to name a few.

A probabilistic model is one that assigns different probabilities to the outcomes such that their sum (or integral) adds up to 1. The shape, spread, skewness, and peakedness are distinct characteristics of a probabilistic model. Different models are distinguished by varying one or more unknown parameters. Each such parameter can vary in a specific range called parameter-space. There could exist multiple candidate models for a given situation. Each such model has its own parameters, which are unknown variables of the population. Several statistical distributions have a single unknown parameter. Examples are Bernoulli, geometric, Poisson, χ^2, exponential, Rayleigh, and T distributions. Most of the other distributions have two parameters. As examples, the binomial distribution has two parameters n and p; negative binomial distribution has r and p with parameter-space $\Omega_p \in (0, 1)$ is continuous and Ω_n is the set of positive integers. These parameters may capture the shape information for some distributions. The location-and-scale (LaS) distributions are those in which the location information (central tendency) is captured by one parameter and scale (spread and skewness) is captured by another parameter. Most of the LaS distributions are of continuous type. Examples are the general normal, Cauchy, and double-exponential distributions. Maximum likelihood estimates (MLE) of the parameters of LaS distributions have some desirable properties. They are also easy and flexible to fit using available data.

The designer's job is to choose the most appropriate one that can describe different sets of data drawn from a population with minimal error. As shown in later chapters, when the parameter $p \to 1$, the binomial distribution BINO(n, p) approaches negative binomial distribution NBINO(n, p) and Poisson distribution POIS(λ) where $\lambda = nq$. Which among them is the best to choose? This may depend on the sample size or the randomness of data. The choice of a model could also depend on some preset conditions such as the available data at hand, data dynamics, dependency on other variables including time-dependency, cost of data collection, desired accuracy level of the built model, adequacy in different situations, etc. Models may have to be constantly updated for time or location dependent data (as in mobile devices in motion).

1.1.2   USES OF DISCRETE PROBABILISTIC MODELS

Discrete probability models are used in various problems in engineering and applied sciences. Count data are quite often described by discrete probability distributions, unless the counts are always very high. For instance, the binomial model is a good choice if the probability of occurrence of a random phenomena can be assumed to remain reasonably constant during a fixed time interval, and the counts are low (say between 2 and 50). The Poisson model could accommodate very large counts because the event occurrences in a unit interval (λ) can be dilated into smaller or larger intervals using an additional variable (λt for temporal processes; details appear in Chapter 6). A study of discrete distributions is also important for professionals in various fields. Data to be modeled are categorical in several fields like bioinformatics (protein sequences with a 20-letter alphabet), genomics (DNA sequences on 4-letter alphabet), etc. The multinomial distribution is a proper choice in such fields. Tail probabilities of some of the distributions are easy to evaluate due to the close relationships between continuous distributions. The CDF of binomial, negative binomial, and Poisson distributions can be expressed as continuous functions like the incomplete beta and gamma functions. Similarly, the binomial and negative binomial distributions are related to the F distribution, and the Poisson distribution has a close relationship with χ^2 distribution. These are discussed in various upcoming chapters. Another important property that is discussed in subsequent chapters is the additivity of independent distributions belonging to the same family. Sometimes, discrete models could provide reliable accuracy for some continuous models.

1.2   DISCRETE RANDOM VARIABLES

A real-valued function defined on the sample space (denoted by Greek letter Ω with or without a subscript that denotes the random variable defined on it) of a random experiment is called a random variable.

Definition 1.1   A random variable is a set function that assigns numerical value to all possible outcomes in the sample space of a random experiment.

By convention, random variables are denoted by uppercase English letters (X, Y, etc.), and particular values by lowercase letters (x, y, etc.). The $\sim$ sign is sometimes used with distributions specified by name and parameters as a symbolic abbreviation for "distributed as." Similarly, IID denotes independently and identically distributed random variables. Thus, $X \sim \text{BER}(p)$ is read as "X is distributed as" Bernoulli distribution with parameter p. Random variables can be discrete or continuous. Discrete random variables assume only discrete values in a finite or infinite range.

Definition 1.2   A random variable X is *discrete* if the sample space Ω_X is finite or countably infinite.

In other words, a random variable that takes a countable number of possible values in a finite or infinite interval is called a discrete random variable. Here, *countably infinite* means that the outcomes in Ω can be rearranged such that a one-to-one correspondence exists with it and the positive integers. As examples, the binomial distribution takes values $x = 0, 1, 2, \ldots, n$ where n is a positive integer > 1, and the Poisson distribution takes values $x = 0, 1, 2, \ldots, \infty$. The possible values of x are assumed to be ordered sequentially in increasing order so that the PMF can be defined for an arbitrary value of x. This is especially important when the random variable assumes a very small number of values (say $x = -2, -1, 0, 1, 2$).

These probabilities can be specified either numerically or using mathematical functions. For example, $f(x) = (2x - 1)/9$ for $x = 1, 2, 3$ is a probability distribution. Probability distributions with unknown parameters are more popular because it can be applied in a variety of practical problems for various values of the parameters. Probability distributions can also be specified empirically, without any unknowns.

Definition 1.3   The probability distribution of a discrete random variable (say X) is the list of all possible values of X along with their respective probabilities in such a way that the probabilities always add up to one.

Some authors use the notation $X(\omega)$ where $X()$ acts like a set-function and ω is an arbitrary outcome of the random experiment.

The values assumed are positive ($x \geq 1$) or non-negative ($x \geq 0$) integers that are equispaced in most of the applications. Theoretically, this is not a restriction. If an employer allows an employee to take either a half-day leave or a full-day leave only, the variable of interest takes integer or half-integer values. Similarly, if kids are charged half of adult charges (as in transportation systems, sports and games events, sky resorts, exhibitions, etc.), the total charges levied from a group is a multiple of an integer or half-integer. As an example in medical sciences, consider the dosage of a medicine to patients of various age-groups. If the adult dose is assumed to be one, dosages for other age-groups can be expressed as fixed fractions of adult dose as *half-dose, quarter-dose*, etc. This is precisely the reason why most pharma companies put a "half-line" on medicines in tablet form where patients can split it (e.g., a 40 mg tablet when split on its half-line will give you two 20 mg portions for the same age-group), although this is not advisable for combo-medicines.[1] If a random variable X takes only integer or half-integer values, a transformation $Y = 2X$ results in an integer valued random variable. However, the majority of discrete distributions are defined on "counts" or "occurrences" that take non-negative integer values $(0, 1, 2, \ldots)$. Continuous random variables on the other hand can assume a continuum of values in an interval.

1.3   CONDITIONAL MODELS

These are obtained by conditioning on one or more events. Let X denote a discrete random variable and E denote an event. If $\Pr(E) > 0$, then $\Pr_{X|E}(x = k) = \Pr[(x = k) \cap E]/\Pr(E)$. From this the conditional expectation follows as $E[X|E] = \sum_x x \Pr[X = x|E]$. If $E_1, E_2, \ldots, E_n$ is a proper partition of the sample space Ω, we have $E(X) = \sum_{j=1}^{n} E[X|E_j] \Pr[E_j]$. If E denotes the occurrence of another dependent random variable Y, $\Pr_{X|Y=y}(x = k|y) = \Pr[(x = k, Y = y)]/\Pr(Y = y)$. The conditional expectation then becomes $E[X|Y = y] = \sum_x x \Pr[X = x|Y]$. The conditional expectation can be used to get the conditional averaging identity $E(x) = E[E(X|N)]$ where N is the conditioning variable. This is used to find the mean of conditional distributions in Chapters 5 and 6.

---

[1]All adult medicines cannot be split to create dosages for children because the concentration, absorption, and side effects can vary considerably among adult and child dosages.

1.4   LINEAR COMBINATION OF RANDOM VARIABLES

A linear combination of random variables is encountered in many branches of engineering. The simplest one is the convex combination of two random variables. If $F_1(x)$ and $F_2(x)$ are the CDFs of two random variables X_1 and X_2, then $G = cF_1(x) + (1-c)F_2(x)$ where c is a nonzero constant in $[0, 1]$ is also the CDF of a random variable. This is simply the CDF of the convex combination $Y = cX_1 + (1-c)X_2$. The mean and variance are, respectively, $c\,E(X_1) + (1-c)E(X_2)$ and $c^2\,V(X_1) + (1-c)^2 V(X_2) + 2c(1-c)\mathrm{COV}(X_1, X_2)$. In the particular case when X_1 and X_2 have the same mean, the mean of Y is $c\,\mu + (1-c)\mu = \mu$. When X_1 and X_2 are independent, $\mathrm{COV}(X_1, X_2) = 0$.

1.4.1   RANDOM SUMS OF RANDOM VARIABLES

In some applications, we may come across a linear combination of random variables in which the number of terms is decided by another random variable. A simple example is the sum of N IID Poisson random variables $Y = X_1 + X_1 + \cdots + X_N$ where N itself is distributed according to another law (say $N \sim \mathrm{BINO}(m, p)$).

```
E(Y) = E(E(Y|N))
V(Y) = V(E(Y|N)) + E(V(Y|N))
```

Consider an insurance company that receives a random number of multiple claims (claims for different categories like medicines, medical procedures like lab tests, scans and x-rays, surgical operations, multiple clinic visits, etc., in medical insurance). Each person claims a random amount of money for one or more categories. This results in a random number of policy holders asking for a sum of a random number of terms. Let $Y = X_1 + X_1 + \cdots + X_N$ denote the total sum to be dispersed. This can be formulated in terms of probability generating function (PGF) as follows. If $P_x(t)$ is the PGF of X_i, then

$$P_y(t) = P_{x_1}(t) * P_{x_2}(t) * \cdots * P_{x_N}(t). \tag{1.1}$$

1.4.2   MIXED RANDOM VARIABLES

The probability associated with a random variable can be obtained as the product of two or more distinct distributions. This book considers only mixed random variables involving two components only. Mixing by three or more random variables is straightforward. The 'mixed-in' components can be discrete or continuous. This results in a new discrete distribution when both of the components are discrete.

Definition 1.4   Definition of mixed random variable.   A random variable in which the associated probability is obtained by mixing (multiplying) the probabilities of two or more independent components is called a mixed random variable.

One example is the noncentral negative binomial (NNB) distribution, which is a discrete mixture of Poisson and negative binomial distributions [50], [59]. As both components are discrete with the same domain $(0, 1, 2, \ldots, \infty)$, the NNB is a discrete distribution. The mixed distribution is of continuous type when one of the components is continuous. Examples are noncentral chi-squared (mixture of Poisson and central chi-squared distribution), noncentral beta (mixture of Poisson and central beta distribution), noncentral F (mixture of Poisson and central F distribution), etc. All of these are continuous distributions with range depending on the range of the continuous component.

1.4.3   DISPLACED RANDOM VARIABLES

Displaced random variables are those in which the lower or upper limit get "displaced" due to a change of origin transformation, but the range remains the same. The lower limit gets shifted (usually to the right) for "displaced" and "truncated" random variables. The difference between them is that the PMF of truncated distributions are obtained by dividing the standard PMF by one minus the truncated (left-out) probability sum. As an example, the zero-truncated Poisson (ZTP) distribution has PMF

$$f(x, \lambda) = \exp(-\lambda)\lambda^x/[(1 - \exp(-\lambda)) \, x!], \quad \text{for} \quad x = 1, 2, 3, \ldots. \tag{1.2}$$

because the probability for $x = 0$ is $\exp(-\lambda)$. Upper-truncated and both-tails truncated distributions are also popular. If truncation occurs at c in the left tail and d in the right-tail, the PMF is given by

$$g(x) = f(x)/(F(d) - F(c - 1)), \quad \text{for} \quad c \leq x \leq d. \tag{1.3}$$

Although the truncation point is in the extreme tails for majority of applications, there are some applications where truncation occurs toward the middle (mean). One example is in size by weight distribution of hazardous waste shredding using hammermill shredders that results in plastic, paper, glass, metal, and sand particles. All of the classical random variables discussed in subsequent chapters can be displaced or truncated by a finite amount. Theoretically, displaced distributions are those obtained by a change of origin transformation $(Y = X \pm c)$. The constant c is assumed to be a nonzero integer for discrete distributions, and a real number for continuous distributions. Left-truncated distributions are special displacements in which the starting value is offset by a positive integer.

Definition 1.5   Definition of displaced random variable.   A random variable in which all the probabilities are displaced (moved) to the left or right is called a displaced random variable.

1.5   DOMAIN OF A RANDOM VARIABLE

The set of values that the random variable takes is called the domain. This will be denoted by lowercase letters like x or y. Unless otherwise stated, the domain is assumed to be the set of equispaced integers. The domain of X can be finite (as in binomial, discrete uniform, hypergeometric, and beta-binomial distributions) or infinite (as in Poisson, geometric, negative binomial, and logarithmic distributions). We naturally expect the probabilities to tail-off to zero beyond a cutoff for distributions with infinite range. Discrete distributions with finite range are more popular in practical applications, while those with infinite range are more important theoretically. This is because some statistical distributions with finite range asymptotically converge to discrete distributions with infinite range as shown below. Positive random variables are those that take only non-negative values. Most of the distributions discussed in this book belongs to this category. Degenerate distribution is a special case in which the random variable of interest takes just one value. Thus, $f(x) = p$ for $x = k$ and $(1 - p)$ otherwise is a degenerate distribution.

The domain differs in various fields. They are strings from a finite alphabet of residues in bioinformatics and genomics. They are usually 4 nucleotides in genomics and 20 amino acids in bioinformatics.

1.6   CDF AND SF

The sum of the left-tail probabilities (up to a fixed point) is called the cumulative distribution functions (CDF). It is an increasing function, upper-bounded by 1.

The CDF of discrete random variables are step functions. Symbolically, $F(x) = \Pr(X \leq x)$ is called the CDF. The PMF can be obtained from the CDF as $f(x) = F(x) - F(x - 1)$. This can also be written as $f(x) = (1 - F(x - 1)) - (1 - F(x))$ because the $+1$ and -1 cancels out. As $(1 - F(x)) = S(x)$, this becomes $f(x) = S(x - 1) - S(x)$. These expressions are used in Chapters 3 and 4. In the case of continuous distributions, this becomes $F(x) = \int_{ll}^{x} f(t)dt$ where ll is the lower limit. The CDF is known by other names in engineering fields. For example, it is called failure distribution function, time-to-failure (TTF) or unreliability function in reliability engineering,[2] where time (continuous) is the variable. The number of repair cycles before the failure of a device can be modeled using a discrete random variable (see Chapter 6). The complement of this is called survival function (SF), confidence level, reliability function, or cumulative tail function as $S(x) = 1 - F(x) = \Pr[X > x]$ (some authors use the R(t) notation where t denotes time). It is called exceedance probability curve in earthquake engineering, disaster management, epidemiology, etc. If c and d are fixed constants such that $c < d$, then $\Pr[c \leq x \leq d] = F(d) - F(c - 1)$ for discrete distributions. As a particular case, the PMF at $x = k$ is given by $F(k) - F(k - 1)$. The CDF of binomial, negative binomial, and Poisson dis-

---

[2]The mean time to failure (MTTF) is $M = \int_{0}^{\infty} t f(t)dt = \int_{0}^{\infty} S(t)dt$ where t denotes time, and $S(t)$ is the survival function (usually of an unrepairable device).

tributions can be expressed as continuous functions like the incomplete beta and gamma functions as shown below. The inverse of the CDF is denoted as $x = F^{-1}(c)$ where the cumulative probability up to x is c (or 100 c% of the total area lies to the left of x). Thus, the median is given by $F^{-1}(0.5)$ or equivalently $S^{-1}(0.5)$. This is more popular for continuous distributions.

The hazard rate and cumulative hazard functions used in reliability engineering are defined in terms of CDF and SF as follows: $h(t) = f(t)/R(t) = f(t)/[1 - F(t)]$ where $f(x)$ is the PMF and $F(x)$ is the CDF. Cumulative hazard functions is $H(x) = \sum_{ll}^{x} h(x)$ where ll is the lower limit. In the continuous case this is given by $H(x) = \int_{0}^{x} f(t)/[1 - F(t)]dt = -\ln[1 - F(x)]$. We could also define a conditional reliability function that gives the probability of survival of a device, given that it has already survived up to a specified time as $c(x) = S(t + x)/S(t)$.

1.6.1   PROPERTIES OF CDF

The CDF satisfies the following properties: (i) $F(\infty) = F(ul) = 1$ where ul denotes upper limit (if not $+\infty$), (ii) $F(-\infty) = F(ll) = 0$ where ll denotes lower limit (if not $-\infty$), (iii) $F(x)$ is monotonically increasing (it is a step function for discrete distributions), (iv) $p(x) = F(x) - F(x - 1)$ and $p(ll) = F(ll)$ (thus $p(0) = F(0)$ for non-negative distributions), and (v) $F(x+) = F(x) \forall x (F(x)$ is a right-continuous function). Conditions (i) and (ii) imply that $0 \leq F(x) \leq 1$. If $F()$ and $G()$ are distribution functions, then $\alpha F(x) + (1 - \alpha)G(x)$ is also a distribution function. Another important property that relates the sum of the CDF with the mean deviation (MD) is given in section 1.8, page 11.

Example 1.6   Telescopic series   Check whether the function $f(x) = 1/[x(x + 1)]$ for $x = 1, 2, \ldots$ is a well-defined distribution.

Solution: Split the PMF using partial fractions as $1/x(x + 1) = 1/x - 1/(x + 1) > 0$ for all x. This being a telescopic series, alternate terms cancel out when it is summed over the range, giving a sum of 1. Hence, it is a distribution. But the mean does not exist because $E(x) = \sum 1/(x + 1)$ is divergent.

1.7   BINOMIAL THEOREM

The binomial theorem with positive and negative exponents has many applications in statistical distribution theory. This section provides an overview of this theorem, which will be used in the sequel. We first consider an expansion for integer powers of a sum or difference of two quantities. More specifically, if n is a positive integer, and x and y are nonzero real numbers, the power $(x + y)^n$ can be expressed as a sum of $n + 1$ quantities in either of two ways as follows:

$$(x + y)^n = \sum_{k=0}^{n} \binom{n}{k} x^k y^{n-k} = \sum_{k=0}^{n} \binom{n}{k} y^k x^{n-k}, \tag{1.4}$$

where $\binom{n}{k}$ denotes $n!/(k!(n-k)!)$. This is most easily proved by induction on n. The numbers $\binom{n}{k}$ (also denoted as $_nC_k$, $C_{n,k}$, etc.) are called *binomial coefficients*, which are always integers when n and k are integers. The special case $\binom{n}{0}$ is defined to be 1 by convention. In the particular case when $x = y = 1$, the above becomes $2^n = \sum_{k=0}^{n} \binom{n}{k}$. As $\binom{n}{k} = \binom{n}{n-k}$, the coefficients in the above expansion are symmetric (hence $2^n = \sum_{k=0}^{n} \binom{n}{n-k}$), which follows by summing in reverse). If n is odd, there are $(n+1)$ terms with $(n+1)/2$ coefficients symmetrically placed. For instance, if $n = 5$, there are 3 coefficients $\binom{5}{0} = \binom{5}{5} = 1, \binom{5}{1} = \binom{5}{4} = 5, \binom{5}{2} = \binom{5}{3} = 10$. If n is even, there are $n/2$ symmetric coefficients with a unique middle coefficient $\binom{n}{n/2}$.

If y is negative, we write $x - y$ as $x + (-y)$ and the above expansion gives

$$(x - y)^n = \sum_{k=0}^{n} \binom{n}{k} x^k (-y)^{n-k} = \sum_{k=0}^{n} \binom{n}{k} (-y)^k x^{n-k} = \sum_{k=0}^{n} \binom{n}{k} (-1)^k y^k x^{n-k}. \quad (1.5)$$

When the index n in the above expansion is negative, we get an infinite series as given below:

$$(x + y)^{-n} = \sum_{k=0}^{\infty} \binom{n+k-1}{k} (-x)^k y^{n-k} = \sum_{k=0}^{\infty} \binom{-n}{k} x^k y^{n-k}. \quad (1.6)$$

In the particular case when $y = 1$ we get

$$(1 + x)^{-n} = \binom{-n}{0} + \binom{-n}{1} x + \binom{-n}{2} x^2 + \cdots = \sum_{k=0}^{\infty} \binom{n+k-1}{k} (-x)^k. \quad (1.7)$$

1.7.1   RECURRENCE RELATION FOR BINOMIAL COEFFICIENTS

Binomial coefficients satisfy many recurrences. These are most often proved by "combinatorial arguments" because $\binom{n}{r}$ denotes the number of ways of choosing r objects from among n objects without replacement. We give only the simplest and most popular recurrences as follows:

1. $\binom{n}{r} = \binom{n-1}{r} + \binom{n-1}{r-1}$.
   This is known as Pascal's identity. As the only arithmetic operation involved is addition, this always returns an integer result.

2. $\binom{n}{r} = \frac{n}{r} \binom{n-1}{r-1}$.
   This recurrence simultaneously decrements both the arguments, and is useful in computing the coefficients for small r values. It is used in subsequent chapters. This could result in approximations due to truncation error resulting from (n/r). A remedy is suggested in next page.

3. $\binom{n}{r} = \frac{n-r+1}{r} \binom{n}{r-1}$.
   This form is useful when n is large and r is small.

4. $\binom{n}{r} = \frac{n}{n-r}\binom{n-1}{r}$.

   This form is useful when n is very large, and r is close to n, so that the decrementing of n is continued until n becomes r. It is used to simplify the MD of binomial distribution.

5. $\binom{-n}{r} = (-1)^r\binom{r+n-1}{r}$.

   This is used in negative binomial and negative hypergeometric distributions.

6. $\binom{n}{r}\binom{r}{m} = \binom{n}{m}\binom{n-m}{r-m}$.

   This form is useful when n and m are large and close-by.

7. $\sum_{r=m}^{n}\binom{r}{m} = \binom{n+1}{m+1}$.

   This combines multiple summations of combinations into a single combination. It is used in finding factorial moments.

8. $\sum_{k\leq n}\binom{r}{k}\binom{s}{n-k} = \binom{r+s}{n}$.

   This is called Vandermonde convolution. It is used in deriving factorial moments of some discrete distributions of hypergeometric family.

Binomial coefficients evaluated by a computing device can sometimes result in approximations. For instance, $\binom{5}{3}$ when evaluated by $\binom{n}{r} = \frac{n}{r} * \binom{n-1}{r-1}$ gives $(5/3) * (4/2) * \binom{3}{1} = 1.66666666 * 2 * 3 = 9.99999999$ due to truncation error. This is because the expression inside the bracket is forcibly evaluated (technically, the division is first evaluated, followed by multiplication in L to R order). If the order of evaluation is modified as $\binom{n}{r} = n * \binom{n-1}{r-1}/r$ without parenthesis, we will get the correct integer result. This argument also applies to options (3) and (4) given above. Alternatively, use Pascal's identity (1).

1.7.2   DISTRIBUTIONS OBTAINABLE FROM BINOMIAL THEOREM

The binomial expansion of the form $(x + y)^n$ for $n > 2$ first appeared in the works of Isaac Newton in 1676. A proof for general n was given by Jacob Bernoulli (1654–1705). There are many statistical distributions that can be derived from the above form of the binomial theorem. Taking $n = 1, x = p$, and $y = q = 1 - p$, we get the Bernoulli distribution (Chapter 2). Setting $n > 1$ to be an integer, $x = p$ and $y = q = 1 - p$, results in the Binomial distribution (Chapter 2). Putting $n = -1, p = -P, q = Q$, we get $f(x) = 1/(Q - P)$, which is a special case of discrete uniform distribution. Setting $n = -m, x = p$, and $y = q = 1 - p$ results in the negative binomial distribution (Chapter 5). Writing $(x + y)^n$ as $x^n(1 + y/x)^n$, putting $y/x = -Q, 1/x = P$, and $n = -1$ we get the geometric distribution (which has infinite range). Put $n = -1$, and write $(x - y)^{-1}$ as $1/x(1 - y/x)^{-1}$. Put $x = (1 + \lambda), y = \lambda$ to get Polya's distribution. Setting $y/x = \theta$, taking logarithm and expanding using $-\log(1 - x) = x + x^2/2 + x^3/3 + \cdots$ results in logarithmic series distribution. Write $(1 + y)^n = (1 + y)^{n_1} * (1 + y)^{n_2}$ where $n_1 + n_2 = n$. Expand each one using binomial theorem, equate identical coefficients on both sides, and divide

RHS by LHS constant to get the hypergeometric distribution (HGD) if n is a positive integer and negative HGD if n is a negative integer.

1.8   MEAN DEVIATION OF DISCRETE DISTRIBUTIONS

The MD is also called the mean absolute deviation or L_1-norm. It is closely associated with the Lorenz curve used in econometrics, Gini index, and Pietra ratio used in economics and finance, and in reliability engineering.

Theorem 1.7   Power method to find the Mean Deviation.
The MD *of any discrete distribution can be expressed in terms of the* CDF *as*

$$\text{MD} = 2 \sum_{x=ll}^{\lfloor \mu-1 \rfloor} F(x), \tag{1.8}$$

where ll *is the lower limit of the distribution,* μ *the arithmetic mean, and* $F(x)$ *the* CDF.

Proof. By definition

$$E|X - \mu| = \sum_{x=ll}^{ul} |x - \mu| p(x), \tag{1.9}$$

where ll is the lower and ul is the upper limit of the distribution. Split the range of summation from ll to $\lfloor \mu \rfloor$ and $\lfloor \mu \rfloor + 1 = \lceil \mu \rceil$ to ul, and note that $|X - \mu| = \mu - X$ for $x < \mu$. This gives

$$E|X - \mu| = \sum_{x=ll}^{\lfloor \mu \rfloor} (\mu - x) p(x) + \sum_{x=\lfloor \mu \rfloor+1}^{ul} (x - \mu) p(x). \tag{1.10}$$

As $E(X) = \mu$, we can write $E(X - \mu) = 0$, where $E()$ is the expectation operator. Expand $E(X - \mu)$ as

$$E(X - \mu) = \sum_{x=ll}^{ul} (x - \mu) p(x) = 0. \tag{1.11}$$

As done earlier, split the range of summation from ll to $\lfloor \mu \rfloor$ and $\lfloor \mu \rfloor + 1$ to ul to get

$$E(X - \mu) = \sum_{x=ll}^{\lfloor \mu \rfloor} (x - \mu) p(x) + \sum_{x=\lfloor \mu \rfloor+1}^{ul} (x - \mu) p(x) = 0. \tag{1.12}$$

Substitute $\sum_{x=\lfloor \mu \rfloor+1}^{ul} (x - \mu) p(x) = -\sum_{x=ll}^{\lfloor \mu \rfloor} (x - \mu) p(x)$ in (1.10) to get

$$E|X - \mu| = \sum_{x=ll}^{\lfloor \mu \rfloor} (\mu - x) p(x) - \sum_{x=ll}^{\lfloor \mu \rfloor} (x - \mu) p(x) = 2 \sum_{x=ll}^{\lfloor \mu \rfloor} (\mu - x) p(x). \tag{1.13}$$

Write (1.13) as two summations

$$E|X - \mu| = 2 \left(\sum_{x=ll}^{\lfloor \mu \rfloor} \sum_{i=ll}^{x} p(i) \right) \qquad (1.14)$$

and substitute $\sum_{i=ll}^{x} p(i) = F(x)$ to get the final result as

$$\text{MD} = 2 \sum_{x=ll}^{\lfloor \mu \rfloor} F(x). \qquad (1.15)$$

The above result is derived under the assumption that the mean is either an integer or a fraction. The summation is carried out to $\mu - 1$ if the mean is an integer. If the mean μ is a non-integer (with a fractional part), let $c = \lfloor \mu \rfloor$ and $\delta = \mu - \lfloor \mu \rfloor$. Obviously, $0 < \delta < 1$. Then a correction term $2\delta F(c)$ must be added to get the correct MD when μ is non-integer. If μ is a half-integer, $\delta = 0.5$ so that $2\delta F(c) = F(c)$. If X is integer-valued, we can write the above as

$$\text{MD} = \sum_{x=ll}^{\lfloor \mu \rfloor} F(x) + \sum_{x=\lfloor \mu \rfloor + 1}^{ul} S(x), \qquad (1.16)$$

where $S(x)$ is the survival function. The above could also be evaluated as

$$\text{MD} = 2 \sum_{x=\lfloor \mu \rfloor + 1}^{ul} S(x). \qquad (1.17)$$

This can be stated as follows:

$$\text{MD} = \begin{cases} 2 \sum_{x=ll}^{\mu-1} F(x) = 2 \sum_{x=\mu+1}^{ul} S(x) & \text{if} & \mu \text{ is an integer,} \\ 2 \left(\sum_{x=ll}^{c} F(x) + \delta F(c) \right) & c = \lfloor \mu \rfloor, \delta = \mu - c & \mu \text{ non-integer,} \\ 2 \left(\sum_{x=c+1}^{ul} S(x) - \delta S(c) \right) & \text{if} & \mu \text{ non-integer.} \end{cases}$$

The above theorem can be easily extended to find the MD of bivariate and multivariate discrete distributions. There is another way to find the MD using mean deviation generating function (MDGF). The MDGF fetches a single coefficient of $t^{\mu-1}$ in the power series expansion of $2P_x(t)/(1 - t)^2$, where $P_x(t)$ is the probability generating function (Chattamvelli and Shanmugam (2019)) [18]. □

1.8.1   NEW MEASURES OF VARIABILITY

The above result can be used to come up with new dispersion measures. As the MD is twice the "sum of the sum" of tail probabilities up to the mean, it is ideal for symmetric distributions. For finite skewed distributions, suppose we take this "sum of the sum" of tail probabilities up to the mean from both the tails as follows:

$$
\begin{cases}
\alpha \sum_{x=ll}^{\mu-1} F(x) + (1-\alpha) \sum_{x=\mu+1}^{ul} S(x) & \text{if} \qquad\qquad \mu \text{ is an integer,} \\
\alpha \left(\sum_{x=ll}^{c} F(x) + \delta F(c) \right) \\
\quad + (1-\alpha) \left(\sum_{x=c+1}^{ul} S(x) - \delta S(c) \right) & c = \lfloor \mu \rfloor,\ \delta = \mu - c \quad \mu \text{ non-integer.}
\end{cases}
$$

This sum can be considered as a new measure of dispersion for $0 < \alpha < 1$. This coincides with twice the MD for $\alpha = 1/2$. It captures the spread in both tails for asymmetric distributions.

1.8.2   A NEW FAMILY OF DISTRIBUTIONS

The result for MD given above can be used to form a new family of distributions. First consider the case where μ is an integer. Rewrite equation (1.15) as

$$
(2/\text{MD}) \sum_{x=ll}^{\mu-1} F(x) = 1, \tag{1.18}
$$

where μ is assumed to be greater than 1. Now define a new distribution

$$
g_1(x; \text{MD}, \mu) = (2/\text{MD}) F(x), \quad \text{for} \quad x = ll,\ ll+1, \ldots, \lfloor \mu \rfloor, \tag{1.19}
$$

where ll is the lower limit of the distribution. As $F(x)$ is monotonically increasing function, the probabilities steadily increase, so that the mode is at $x = \mu - 1$. If $G_1(x)$ is the CDF, μ_1 is the mean and MD1 the mean deviation of this new distribution, we could define another new distribution $g_2(x; \text{MD1}, \mu_1) = (2/\text{MD1}) G_1(x_i)$, and so on. Proceed as above with the correction term when the mean is not an integer.

Example 1.8   Variance of discrete distribution as SF   Prove that the variance of discrete distributions can be expressed in terms of tail probabilities.

Solution: We know that the MD is an L_1-norm, and σ^2 is an L_2-norm. We found above that $\text{MD} = 2 \sum_{x=ll}^{\lfloor \mu \rfloor} F(x) = 2 \sum_{x=\lfloor \mu \rfloor+1}^{ul} S(x)$. A linear relation exists among the MD and variance for most discrete distributions [37], [41] as $\text{MD} = c\mu_2\, f_m$, where c is a constant, $\mu_2 = \sigma^2$, and f_m is the PMF evaluated at the integer part of the mean $m = \lfloor \mu \rfloor$. This gives $c\mu_2\, f_m = 2 \sum_{x=ll}^{\lfloor \mu \rfloor} F(x)$. Divide both sides by $c f_m$ to get

$$
\sigma^2 = (2/(c*f_m)) \sum_{x=ll}^{\lfloor \mu \rfloor} F(x) = (2/(c*f_m)) \sum_{x=\lfloor \mu \rfloor+1}^{ul} S(x). \tag{1.20}
$$

When $m = \mu$ is a non-integer, the correction term mentioned above must be applied.

Table 1.1: Mean deviation of binomial distribution using our power method (Equation 1.17) for np a half-integer

$n, p\,(np)$	0	1	2	3	4	Eq. (1.17)	Final
15, 0.3 (4.5)	0.0047	0.0353	0.1268	0.2969	0.5155	0.92742	1.442913
18, 0.25 (4.5)	0.0056	0.0395	0.1353	0.3057	0.5187	1.39742	1.91171
8, 0.8125 (6.5)	0.0009	0.0078	0.0455	0.1762	0.4594	0.46098	0.920416

First column gives n, p (np) values of binomial. Second column onward are the values computed using Equation (1.17). Seventh column gives the uncorrected MD using Equation (1.15). Correction term $2\delta F(c)$, where $c = \lfloor \mu \rfloor$ and $\delta = \mu - \lfloor \mu \rfloor$ is added to get the correct MD in the last column.

1.9   TRUNCATED MODELS

Truncated distributions were briefly introduced in §1.4.3 in page 6. Zero-truncated distributions (ZTDs) have PMF of the form $g(x; \mu) = f(x; \mu)/[1 - F(0)], x = 1, 2, \ldots,$ where μ denotes unknown parameters and $F(x)$ denotes the CDF. If truncation occurs in the left tail at m, the PMF becomes $g(x; \mu) = f(x; \mu)/[1 - F(m)], x = m + 1, m + 2, \ldots$. As $\sum_{x=0}^{m} f(x; \mu) + \sum_{x>m} f(x; \mu) = 1$, the above can be alternately expressed as $g(x; \mu) = f(x; \mu)/\sum_{x>m} f(x; \mu)$, for $x = m + 1, m + 2, \ldots$. The mean is $\mu = E(X|X > 0)$, which for truncated Poisson distribution is $\lambda/[1 - \exp(-\lambda)]$. $Var(X|X > 0) = \mu/[1 - f(0)] - f(0)E(X|X > 0)^2$.

1.10   ZERO-INFLATED DISTRIBUTIONS

As the name implies, zero-inflated distributions (ZIDs) are statistical models in which zero value is decoupled from the rest, and weighted by $(1 - \theta)$ where θ is a user chosen constant for the problem at hand. Thus, there exist a degenerate distribution at zero and a ZTD for the rest of the data. In other words, these are dual-process models in which zero is in one state (called zero state) and the rest of them is in another state (normal state). This may sometimes result in a bimodal distribution. As an example from actuarial science, this class may represent people who did not opt for a particular risk so that they are unable to claim. Thus, ZIDs are promising models for over-dispersed data. Contagious events are those in which a positive correlation exists among adjacent event occurrences. Distributions defined on contagious events are often over-dispersed. ZID is an excellent choice to model data that have excess zero values. A suitable choice of θ (some authors use π or $1 - \pi$ instead of θ) can improve statistical fit of the data:

$$f(x;\theta) = \begin{cases} 1-\theta & \text{if } x = 0, \\ \theta g(x;\theta) & \text{if } x > 0, \end{cases}$$

where $g(x;\theta)$ is the ZTD. This shows that a ZID is defined in terms of a ZTD. A Bernoulli distribution $BER(\theta)$ is used to decide which of the two processes is chosen to generate data. It generates more zeros when θ is small, and *vice versa*.

There is one more way to define a ZID. This is also called the misrecorded distribution that justifies errors in capturing zero-class of a variable. Consider a regular PMF $f(x;\mu)$ where μ denotes one or more parameters. Let $p_0, p_1, \ldots$ be the associated probabilities so that $\sum_k p_k = 1$. Decouple the probability p_0 (associated with 0), and move it to zero state

$$f(x;\theta) = \begin{cases} 1-\theta+\theta p_0 & \text{if } x = 0, \\ \theta f(x;\mu) & \text{if } x > 0. \end{cases}$$

The mean is $(1-\theta+\theta p_0)*0 + \theta \sum_{x=1}^{\infty} x * f(x;\mu) = \theta\mu$. The variance can easily be found for those distributions with an $x!$ in the denominator using $V(X) = E(X(X-1)) + E(X) - E(X)^2$.

1.11   HURDLE MODELS

These models get the name because the zero-state is considered as a hurdle. Whether a count variable has a zero value or positive value is checked using a binomial model. The conditional distribution of positive values is as above a zero-truncated model. We say that the hurdle is crossed if the count for zero class as given by the model is positive. This model assumes that there is only one source of zeros count. It finds applications in ecology, business, health economics, psychology, and various other fields.

1.12   POWER SERIES DISTRIBUTIONS

Several discrete distributions discussed in subsequent chapters belong to a class of distributions called power series distributions (PSD). Formally, if a bounded function $f(\theta)$ can be expanded as a power series $f(\theta) = \sum_{k=0}^{\infty} a_k \theta^k$, where a'_ks are constants, we could define a PMF as

$$\Pr[X = x] = a_k \theta^k / f(\theta) \quad \text{for} \quad k = 0, 1, \ldots, \infty, \tag{1.21}$$

examples of which are binomial, Poisson, geometric, and logarithmic series distributions. Differentiating above w.r.t. θ gives $f'(\theta) = \sum_{k=1}^{\infty} k\, a_k \theta^{k-1}$, from which $E(X) = \sum_{k=1}^{\infty} k\, a_k \theta^k / f(\theta) = \theta\, f'(\theta)/f(\theta)$. Similarly, $E[X(X-1)] = \theta^2\, f''(\theta)/f(\theta)$.

1.13   SUMMARY

Essential concepts on random variables are introduced in this chapter. A thorough knowledge on discrete models, domain, CDF, and SF of discrete distributions are essential to understand subsequent chapters. A discussion of binomial theorem for positive and negative indices will make it easier for the reader to appreciate its relationship with various distributions. The chapter ends with a discussion of zero-truncated and zero-inflated models.

CHAPTER 2

Binomial Distribution

After finishing the chapter, readers will be able to . . .

- Understand Bernoulli and binomial distributions.

- Explain properties of binomial distribution.

- Describe special binomial distributions.

- Comprehend applications of binomial distribution in engineering.

- Explore binomial distribution in applied sciences.

2.1 INTRODUCTION

Bernoulli distribution is one of the simplest discrete distributions. A good understanding of this distribution is essential to grasp the binomial, geometric, and negative binomial distributions discussed in subsequent chapters. This distribution is based upon the `Bernoulli trial` defined below. Either a dichotomous categorical variable, also called binary variable (like gender) or a quantitative variable (like income), may underlie the experimentation. Assigning probabilities to the outcomes is much easier in the case of categorical variables.

2.1.1 BERNOULLI TRIALS

A Bernoulli trial is a random experiment that results in exactly two possible outcomes each time it is repeated, which are by convention denoted by "success" (S) and "failure" (F), in which the probability of success is the same every time it is conducted under identical conditions. Such an experiment is known as a *Bernoulli trial*. Here "trial" is synonymous with "experiment" or "experimentation." It may also result from a physical test of a quality, detection of "noise only" or "noise plus signal" in signal processing problems, measurement of an attribute (like body temperature), presence of a symptom or a microbe (like COVID-19), or from accumulated knowledge from past data (like relative frequency of occurrence of an event like defect). These are discussed in detail in numerous examples.

2.1.2   "SUCCESS" DEMYSTIFIED

As mentioned previously, the outcomes of a Bernoulli trial are called "success" and "failure." What does "success" mean? Here the meaning of the word "success" and "failure" should not be taken literally—it simply means two dichotomous outcomes of an experiment. It is the experimenter who decides what "success" and "failure" are. In the coin tossing experiment, one experimenter may call Head as "success" and Tail as "failure," whereas another one may define it the opposite way. These are mutually exclusive and exhaustive to indicate that other possibilities do not exist. Thus, when a coin is tossed, we assume that it does not end up on its edge, but comes to rest on either of the two sides. Similarly, an item inspected may either meet the quality standard or not, a component may be either working or not-working according to a set standard.

The event may be absence or presence of a symptom in medical sciences, infected or non-infected with a disease irrespective of whether all the symptoms are present in epidemiology, stable or unstable habitat, sighted and not-sighted (like birds or animals in a location) in ecology, presence or absence of a plant or species in zoology, fertility of an egg or sperm in biology, presence of a quality within acceptable limits in hydrology, mining engineering, etc., profitable or unprofitable in economics and management, presence or absence of a microbe in a finely sized grid overlaid on a microscopic plate in microbiology, growing or shrinking in investment and business, etc. In information sciences the event may denote whether a login attempt was successful or not, or if a data packet or message was transmitted successfully or not. In engineering fields it can denote faulty or non-faulty, working or defunct, closed or open (as in electrical circuits), detected or undetected (radioactivity, smoke, abnormality, etc). In statistical quality control (SQC), this may denote within permissible limits or outside the control limits. The events may also be based on known characteristics of subjects like diabetes (diabetic, non-diabetic), gender (male, female), smoking habits (smoker, non-smoker), drinking habits (drunkard, tee-to-taller), etc. It represents the presence of a trait (like autism, learning disability, etc.) in education and psychology. The above examples show that either the time, space, volume or a combination of them can be used as a frame-of-reference in forming a Bernoulli process (a discrete random sequence with the same probability of happening that takes only two values, say 0 and 1. It is called a *binary white noise process* when the probability of occurrence is $p = 1/2$ that denotes equally likely occurrence). A binary counting process is realized as the sum of several independent Bernoulli processes where the occurrence probability is assumed to remain steady.

2.1.3   SPATIAL PROCESSES

Processes that occur over the surface of a physical object (microscopic plates, fingerprint and retina scanners) or living organisms (human body, skin of marine species or animals) are called spatial processes. For example, dense bacterias are found on the surface of the skin and small amounts in the neighboring muscles or blood vessels in dermatology, microbes grow on the surface and infects interior regions of perishable foods like fish, meat, and some vegetables. These may be modeled using two or more Bernoulli processes (one for the surface, and others for interior regions), as explained in Chapter 6. Some applications like x-ray, tomography, molecular spectroscopy, radio-telescope, etc. convert 3D data into 2D images. A 2D model built from the resulting data can throw insight into the actual 3D process.

2.1.4 DICHOTOMIZATION USING CUTOFF

Several engineering and scientific fields use a cutoff (or threshold level) to binarize quantitative variable. A simple example is classifying a person as having fever or not using the body temperature. When a device (like a light bulb, a spark-plug of an automobile), or a component like battery of a smartphone partially works, we need to decide a cut-off where it should be considered as working. Consider high winds, earthquakes, or other natural disasters. We may use quartiles (which divide data into four parts using frequency of occurrence) or deciles (which divide data into ten parts) of the intensity to decide whether it is below or above the cutoff. Similarly, capacitance, resistance, or voltage levels are used as cutoff in electronic engineering. Particle-size distribution in geotechnical engineering uses ratio of particles by weight finer to distinguish between poorly graded and well graded particles. The cutoff value may also be based upon a battery of attributes. One example is in insurance and actuarial sciences where a customer is classified as risky using a set of variables like *prior accidents*, *total driving experience*, *education level*, *physical or visual impairments*, etc. In hazardous industries where work-related injuries are common, an injured worker may be classified as "requiring first-aid" or "requiring hospital admission" using the severity of the injury that uses multiple factors. In these cases, below cutoff values are assumed to be zero and above values as one. This makes it into two possible outcomes.

The values taken by a Bernoulli random variable are by convention assumed to be 0 and 1. This is called the *domain* or *support* of the distribution. It is convenient to arrange the values in sorted order (from low to high), although this is not a theoretical requirement. This book follows this convention so that the probability p is associated with the value 1. In other words, $X(\text{success}) = 1$ and $X(\text{failure}) = 0$. This is convenient to denote the non-occurrence ($x = 0$) or occurrence ($x = 1$) of an event. However, nothing precludes an experimenter from defining other domains for a random variable. Some applications in game theory, digital signal processing, cryptography, stochastic processes, and some puzzles assume that the values are -1 and $+1$. For instance, a robotic motion that takes one unit to the right with probability p and one unit to the left with probability $q = 1 - p$ can be modeled using a Bernoulli process. Note that this is just a transformation of the classical Bernoulli random variable as $Y = 2X - 1$, so that $x = 0$ maps to $y = -1$, and $x = 1$ maps to $y = 1$. Similarly, we could map a Bernoulli random variable X to $(-k, +k)$ using the transformation $Y = k(2X - 1)$. As another example, consider a game played between two persons X and Y with respective initial amounts m and n. The loser of each game gives a unit amount to the winner (so that loser losses 1 point and winner gains 1 point).

The states of a player after each game can be modeled as a Bernoulli process. If X denotes the number of games to be played for one of them to go bankrupt, the minimum number of games needed is $\min(m, n)$, and the maximum is ∞. Unless otherwise noted, we will assume that the values are $\{0, 1\}$. This is because other related distributions like binomial, geometric, and negative binomial distributions are easy to model with these values.

2.2   BERNOULLI DISTRIBUTION

This distribution is named after the Swiss mathematician Jacques Bernoulli (1654–1705). The Bernoulli distribution results from a random experiment in which each outcome is either a success (denoted by 1) with probability p, or a failure (denoted by 0) with probability q so that $p + q = 1$ (π is used instead of p in some engineering fields, and θ is sometimes used in SQC and Bayesian statistics). Thus, fixing the value of p automatically fixes the value of q. A question that naturally arises is what should be chosen as p? This is not an issue because p and q are simply place holders for probabilities. The actual values depend upon the research hypothesis. Usually, p is chosen as the probability of a fault in engineering sciences. It is chosen as the probability of the presence of a symptom or condition in medical sciences. Thus, this distribution finds applications in a variety of fields.

2.2.1   BASIC PROPERTIES OF BERNOULLI DISTRIBUTION

The PMF of a Bernoulli random variable is

$$f(x; p) = p^x q^{(1-x)}, \quad x = 0 \text{ or } 1, \text{ and } 0 \le q = 1 - p \le 1. \tag{2.1}$$

The Bernoulli distribution is denoted by $\mathrm{BER}(p)$, with a single parameter p. This could also be expressed in the following convenient form:

$$f(x; p) = \begin{cases} p^x(1-p)^{1-x} & \text{for } x = 0, 1; \\ 0 & \text{otherwise.} \end{cases}$$

The $p = 0$ or $p = 1$ cases are called degenerate cases, as there is no randomness involved.

The mean and variance of a Bernoulli distribution are $\mu = p$, $\sigma^2 = pq$, respectively. As the only values of x are 0 and 1, we get the mean as $E(X) = 0 * q + 1 * p = p$. Similarly, $E(X^2) = 0^2 * q + 1^2 * p = p$, so that the variance is $E(X^2) - E(X)^2 = p - p^2 = p(1 - p) = pq$. The coefficients of skewness and kurtosis are, respectively, $\beta_1 = (1 - 2p)/\sqrt{(pq)}$, $\beta_2 = 3 + (1 - 6p)/pq$. The coefficients of variation (CV) is a measure of the variability in the data as a percentage of its mean value. It is defined for the population as σ/μ, and for a sample as $s/\overline{x}$. It is useful to compare variability of two data sets with means far apart. For example, the recovery period for COVID-19 patients differ significantly among seniles and youth. Applied scientists prefer CV to compare such variations. That is the reason why "financial risk" in finance and stock markets is measured using CV. For BER(p) it is $(pq)^{1/2}/p = \sqrt{q/p}$. A detailed investigation of CV for various distributions appears in Shanmugam and Singh (2003) [97].

If we observe k successes in n independent Bernoulli trials, an estimate of p is obtained as $p = k/n$. The PGF is easily obtained as $P_X(t) = (q + pt) = 1 - p(1 - t)$, and characteristic function is $\phi(t) = q + p \exp it = 1 - p(1 - \exp(it))$. Hence, all moments about zero are p.

2.2.2   RELATED DISTRIBUTIONS

There are many other probability distributions based upon Bernoulli trials. For example, the binomial, negative binomial, and geometric distributions mentioned above; success-run distributions are all defined in terms of independent Bernoulli trials. Binomial distribution described below results as the sum of n independent Bernoulli trials. Dependent Bernoulli trials (where the probability of success could vary from trial to trial) results in Poisson-Binomial distribution. Geometric distribution arises from independent Bernoulli trials until a success occurs. The negative binomial distribution is an extension of geometric distribution in which independent Bernoulli trials are carried out such that the nth trial results in rth success. Hypergeometric distribution results when random sampling is done without replacement from a finite dichotomous population.

Example 2.1   CDF of Bernoulli distribution   Suppose p denotes the probability of a trait in a group of persons. Define a random variable X that takes the value 1 if trait is present and is 0 otherwise. Find the PMF and CDF of X.

Solution: Assign a random variable X to the two possible outcomes as $P(\text{trait}) = p$ and $P(\text{trait not present}) = q = 1 - p$. The PMF is expressed as

$$f(x) = \begin{cases} q = 1 - p & \text{if } x = 0 \quad (\text{trait not present}) \\ p & \text{if } x = 1 \quad (\text{trait}). \end{cases}$$

As there are only two possible values, the CDF is obtained as

$$F(x) = \begin{cases} q = 1 - p & \text{if } x = 0 \quad (\text{trait not present}) \\ 1 & \text{if } x = 1 \quad (\text{trait}). \end{cases}$$

The CDF is $(1 - p)^{1-x}$ because $F(0) = (1 - p)$, and $F(1) = 1$. Similarly, SF is $S(x) = p^x$ because $S(1) = p$ and $S(0) = 1$.

The classical Bernoulli distribution uses a single unknown parameter p. Some applications in biology, genomics, and ecology represents this as $p = \theta/(1 + \theta)$ (so that $q = 1/(1 + \theta)$) as

$$f(x; \theta) = \begin{cases} [\theta/(1 + \theta)]^x [1/(1 + \theta)]^{1-x} & \text{for } x = 0, 1; \\ 0 & \text{otherwise.} \end{cases}$$

Table 2.1: Properties of Bernoulli distribution

Property	Expression	Comments
Range of X	$x = 0, 1$	Discrete, finite
Mean	$\mu = p$	
Variance	$\sigma^2 = pq$	$\Rightarrow \mu > \sigma^2$
Skewness	$\gamma_1 = (1 - 2p)/\sqrt{pq}$	$= (q - p)/\sqrt{pq}$
Kurtosis	$\beta_2 = 3 + (1 - 6pq)/pq$	
MD	$2pq$	
Moments	$\mu'_r = p$	
CDF	$(1 - p)^{1-x}$	$= (1 - p)$ for $x = 0$, and 1 for $x = 1$
SF	p^x	$= 1$ for $x = 0$, and p for $x = 1$
MGF	$(q + pe^t) = 1 + p(e^t - 1)$	$= 0.5\,(1 + e^t)$ if $p = q$
PGF	$(q + pt) = 1 + p(t - 1)$	$= 0.5\,(1 + t)$ if $p = q$
CV	$(q/p)^{1/2}$	$\sqrt{q/p}$
Additivity	$\sum_{i=1}^{n} X_i = \mathrm{BINO}(n, p)$	Independent
Bernoulli distribution is the building block of binomial, geometric, negative binomial, and success-run distributions.		

Consider a gene pool of size $2N$ at generation t. Probability that two ancestral lineages coalesce is $p = 1/(2N)$, and not coalescing is $q = 1 - 1/(2N)$. As these are mutually exclusive, it can be modeled by BER(p). If the coalescence is considered from the present generation t to past generations t-k, the probability of a coalescence event is $1/(2N)$ in each of the generations. See Table 2.1 for summary of properties.

Example 2.2   Product of two IID Bernoulli random variables   If X and Y are IID BER(p), find the distribution of $U = X * Y$.

Solution: X and Y both takes the values 1 with probability p, and 0 with probability $q = 1 - p$. Hence, XY takes the value 0 when either or both of X and Y take the value 0 with probability $q^2 + qp + pq$. Here, q^2 is the probability that both of them takes the value 0, and qp, pq are the probabilities that either of them takes value 0 and other takes the value 1. Write $q^2 + qp = q[q + p]$ and use $q + p = 1$ to get q. Next combine q with pq to get $q + pq = q(1 + p)$. Write q as $(1 - p)$ to get $(1 - p) * (1 + p) = 1 - p^2$. Probability that XY takes the value 1 is p^2. Hence, XY has a Bernoulli distribution with probability of success p^2. Symbolically, $XY \sim$

BER(p^2). This can be extended to the case $\prod_{i=1}^{n}(1 + kX_i)$ where k is a constant and X_i are IID BER(p).

2.2.3   RANDOM NUMBERS

The indicator variables used in several applied science fields is a special case of the Bernoulli random variable:

$$I_a(\omega) = \begin{cases} 1 & \text{if} \quad \omega \in (A) \\ 0 & \text{(otherwise)}. \end{cases}$$

It is quite easy to generate random numbers from this distribution using a uniform random number generator in [0,1] range, as follows: Input the value of p in BER(p) in the range $0 < p < 1$. Generate random number u in the range $0 < u < 1$. If $u \leq p$ output 1 else output 0. Repeat this process n times to get a random sample of size n from BER(p).

Problem 2.3   If $X_1, X_2, \ldots, X_n$ are IID BER(p) variables, $S = X_1 + X_2 + \cdots + X_n$, and $W = X_1 + 2X_2 + \cdots + nX_n$, then $Z_n = (W - ((n+1)/2)S)/((n+1)W(n-W)/12)^{1/2} \to N(0,1)$ as $n \to \infty$.

2.3   BINOMIAL DISTRIBUTION

Binomial distribution is a natural extension of Bernoulli distribution for two or more independent trials ($n > 1$). It was first derived in its present form by the Swiss mathematician Jacques Bernoulli which was published posthumously in 1713, although the binomial expansion (for arbitrary n) was studied by Isaac Newton (1676) and Blaise Pascal (1679), among others. It can be interpreted in terms of random trials, or in terms of random variables. Consider n independent Bernoulli trials each with the same probability of success p. We assume that the trials have already occurred. We are interested in knowing how many successes have taken place among the n trials. This number called binomial random variable is any integer from 0 to n, inclusive. Assuming that there are x successes, this can happen at any of the n positions in $\binom{n}{x}$ ways. Here, $\binom{n}{x} = n!/[x!(n-x)!]$ (which is sometimes written as nC_x) is the combination of n things taken x at a time. Expand the $n!$ in the numerator and cancel out with $(n-x)!$ in the denominator to get the alternate form $n_{(x)}/x!$ where $n_{(x)}$ denotes Pochhammer falling factorial notation (Chattamvelli and Shanmugam (2019)) [18].

Since the probability of success remains the same from trial to trial, the probability of x successes and $n - x$ failures is

$$f(x; n, p) = \begin{cases} \binom{n}{x} p^x q^{n-x} & 0 \leq q = 1 - p \leq 1 \quad \text{if} \quad x = 0, 1, 2, \ldots, n \\ 0 & \text{elsewhere}. \end{cases}$$

It is called binomial distribution because it is the xth term in the binomial expansion of $(q + p)^n$, and $\binom{n}{x}$ is called binomial coefficient. An easy way to grasp this is using the binomial expansion $(S + F)^n$ where the coefficients are the rows of Pascal's triangle. For $n = 3$ this becomes

$$(S + F)^3 = SSS + 3\{SSF\} + 3\{SFF\} + FFF.$$

Here SFF denotes the three possible ways to permute one S and two F's, and so on. A question that arises is "What is the distribution of the number of Failures?" This is obtained by exchanging p and q in the PMF because $(p + q)^n = (q + p)^n = \sum_x \binom{n}{x} q^x p^{n-x}$. In this case the unknown parameters are n and q. It can be extended to non-integer values of n, if the expression above is rewritten as:

$$f(x; n, p) = \begin{cases} \Gamma(n + 1)/[x!\Gamma(n - x + 1)]p^x q^{n-x} & 0 \leq q = 1 - p \leq 1 \\ 0 & \text{elsewhere.} \end{cases}$$

It can be represented diagrammatically as a tree in which each node has two branches (called full-binary tree). It is widely used in civil engineering, manufacturing, reliability engineering, high-voltage engineering (HVE), etc. It belongs to the exponential family. To see this, write the PMF as $\binom{n}{x}(p/q)^x q^n$. As $\exp(\log(x)) = x$, this can be written as

$$\exp\left(\log\left(\binom{n}{x}\right) + x \, \log(p/q) + n \, \log q\right). \tag{2.2}$$

In medical and allied fields, we assume that there exist a finite set of individuals $S = \{X_1, X_2, \ldots, X_N\}$. Each of these individuals either possess an attribute Y or does not possess it (has complementary attribute $\bar{Y}$). Binomial distribution results when n of the individuals are drawn randomly in such a way that the probability of the attribute does not change from one draw to another. This happens when either the sampling is with replacement, or when the population size is very large (so that the probability variations are negligible).

Problem 2.4   The probability that a teenager will have a broken bone is 3/1000 per year. If a school has 420 teens, what is the probability that (i) at least one teen will have broken bone, (ii) at most 2 have broken bone, or (iii) none have a broken bone in a year?

In general, it is a suitable choice when a testing problem involves "yes/no" type of answers, and the sample size is small. A linear combination of binomial distributions is used in expectation maximization algorithms (with missing data).

The random variable interpretation of binomial distribution is based upon independent Bernoulli trials. Let $X_1, X_2, \ldots, X_n$ be a sequence of IID Bernoulli random variables with the same parameter p, and the values assumed by each of them are $\{0, 1\}$. Then the sum $X = X_1 + X_2 + \cdots + X_n$ has a binomial distribution with parameters n and p. We denote this distribution as BINO(n, p) and the PMF value at x as BINO(n, p; x) or BINO(x; n, p). As shown below, the mean of this distribution is $\mu = np$. Write this as $p = \mu/n$ and substitute in the PMF to obtain

$$f(x;n,\mu) = \binom{n}{x}(\mu/n)^x(1 - \mu/n)^{n-x}. \tag{2.3}$$

This can alternately be written as

$$f(x;n,\mu) = \binom{n}{x}(\mu/n)^x((n - \mu)/n)^{n-x} \tag{2.4}$$

with mean μ and variance $\mu(n - \mu)/n$.

Next, write the PMF $\binom{n}{x}p^x q^{n-x}$ as $\binom{n}{x}(p/q)^x q^n$. Put $p/q = p/(1 - p) = \theta$. Cross-multiply and solve for p to get $p = \theta/(1 + \theta)$. The PMF now becomes

$$f(x;n,\theta) = \binom{n}{x}[\theta/(1 + \theta)]^x[1/(1 + \theta)]^{n-x}, \quad \theta > 0. \tag{2.5}$$

Now consider a Bernoulli distribution with domain $-1, +1$ as $X_k = +1$ if kth trial is a success with probability p, and -1 otherwise. Let $T = X_1 + X_2 + \cdots + X_n$. The possible values of T are alternate integers in the range from $-n$ to $+n$. For n even, the possible values are $\{-n, -n + 2, \ldots, -2, 0, +2, \ldots, n - 2, n\}$ and for n odd, the possible values are $\{-n, -n + 2, \ldots, -1, +1, \ldots, n - 2, n\}$. This is obtained as $Y = 2X - n$ where X is BINO(n, p). Geometrically, T represents discrete random walks on the integer valued lattice.

Example 2.5 Linear random walks   Suppose a particle is placed at the center of a line marked zero. It takes unit jumps to the right with probability p, and unit jumps to the left with probability $q = 1 - p$. What is the final position of the particle after n jumps if the probability remains the same in each jump?

Solution: This problem is very similar to the above problem. The possible values are integers in the range $\{-n, +n\}$. More jumps to the negative-side will occur when $q > p$ and *vice versa*. Assume that there are r jumps to the right and $n - r$ jumps to the left. Then the final resting position after n jumps is $r - (n - r) = 2r - n = x$ (say). This gives $r = (x + n)/2$. Substitute in the binomial PMF to get the distribution as

$$f(x;n,p) = \binom{n}{(n + x)/2}p^{(n+x)/2}(1 - p)^{(n-x)/2}. \tag{2.6}$$

Table 2.2: Binomial probabilities example

	0	1	2	3	4	5	Sum
p	q^n	$\binom{n}{1}pq^{n-1}$	$\binom{n}{2}p^2q^{n-2}$	$\binom{n}{3}p^3q^{n-3}$	$\binom{n}{4}p^4q^{n-4}$	$\binom{n}{5}p^5q^0$	
0.1	0.59049	0.32805	0.0729	0.0081	0.00045	0.00001	1.0
0.5	0.03125	0.15625	0.3125	0.3125	0.15625	0.03125	1.0
0.9	0.00001	0.00045	0.0081	0.0729	0.32805	0.59049	1.0

2.3.1 DEPENDENT TRIALS

A binomial distribution results from n independent Bernoulli trials ($n \geq 2$). The classical binomial distribution ensue when each of them are independent, and dependent binomial distribution results when at least two of the trials are dependent.

2.4 PROPERTIES OF BINOMIAL DISTRIBUTION

There are two parameters for this distribution, namely the number of trials ($n > 1$; an integer is sometimes called the index or order of the distribution), and the probability of success in each trial (p) which is a real number between 0 and 1.

This probability remains the same from trial to trial, which are independent. This distribution is encountered in sampling with replacement from large populations. If p denotes the probability of observing some characteristic (there are x individuals that have the characteristic in the population, so that $p = x/N$ where N is the population size) the number of individuals in a sample of size n from that population has the characteristic is given by a binomial distribution BINO(N, p).

2.4.1 MOMENTS

As the trials are independent, and $X = X_1 + X_2 + \cdots + X_n$, the mean is $E(X) = p + p + \cdots + p = np$. Similarly, the variance of X is $V(X) = V(X_1) + V(X_2) + \cdots + V(X_n) = pq + pq + \cdots + pq = npq$. Hence, $\mu = np$, $Var(X) = npq = \mu_1 * q$. Thus, the variance is a quadratic in p (i.e., $np(1 - p)$) for binomial distribution. The ratio σ^2/μ called excess dispersion index (EDI) is $npq/np = q$. EDI is used to trace market volatility in financial applications. Note that when $p \to 0$ from above, $q \to 1$ from below, and the variance $\to \mu$. This results in a distribution with the same mean and variance. The symmetry of variance for fixed n indicates that the variance of BINO(n, p) and BINO(n, q) are the same. Method of moments estimators are obtained by equating the expressions for the sample mean and variance to their population counterparts. This gives $\bar{x} = \mu = np$ and $s^2 = npq = \mu q$. Solve simultaneously to get $\hat{q} = s^2/\bar{x}$, so that $\hat{p} = 1 - s^2/\bar{x}$ and $\hat{n} = \bar{x}/\hat{p} = \bar{x}/(1 - s^2/\bar{x})$.

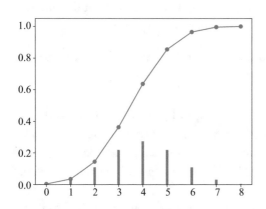

Figure 2.1: PMF and CDF of binomial ($n = 8, p = 0.5$).

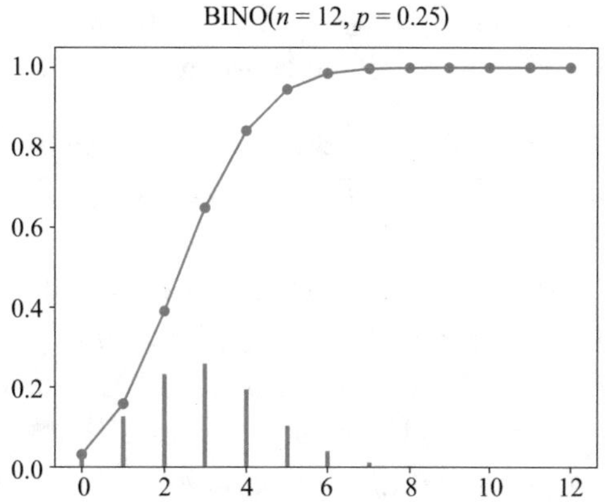

Figure 2.2: PMF and CDF of binomial ($n = 12, p = 0.25$).

Example 2.6   Maximum variance of a binomial distribution   Prove that the maximum variance of a binomial distribution $\text{BINO}(n, p)$ as a function of p is $n/4$.

Solution: We know that the variance is given by $V(n, p) = npq = np - np^2$. Differentiating w.r.t. p we get $\frac{\partial}{\partial p}V(n, p) = n - 2np$. Equating to zero and solving for p gives $p = 1/2$. As the second derivative is $-2n$, this indeed gives the maxima. Substitute in the above to get $V(n, p) = n * (1/2) * (1/2) = n/4$. This can be increased without limit by letting $n \to \infty$.

The ratio of two probabilities of a discrete distribution at distinct ordinal (x) values provides the relative likelihood of the random variable taking a value at one level vs. the other. This is useful to calculate the probabilities recursively, to locate the mode, and to develop moment recurrences [17]. As a special case, the ratio $\Pr[X = 1]/\Pr[X = 0]$ is called the incidence jump rate (IJR), which for the binomial distribution is np/q (Shanmugam and Radhakrishnan (2011) [93]).

Example 2.7   Mode of binomial distribution   Find the mode of the binomial distribution.

Solution: Consider the ratio

$$f(x; n, p)/f(x - 1; n, p) = (n - x + 1)p/(xq). \tag{2.7}$$

Add and subtract xq in the numerator and combine $-xp - xq$ to get $-x$. Then the ratio becomes $1 + [(n + 1)p - x]/(xq)$. The bracketed expression in the numerator is positive when $(n + 1)p > x$ and negative otherwise. If $(n + 1)p = x$, the bracketed expression vanishes giving two modes at $x - 1$ and x. Otherwise the mode is $\lfloor (n + 1)p \rfloor$. This shows that the binomial distribution is uni-modal for all values of p and n, except when $(n + 1)p$ is an integer.

Example 2.8   Find the mean and variance   Find the mean and variance of a distribution defined as $f(-1) = f(+1) = pq$, and $f(0) = p^2 + q^2$.

Solution: The mean is $E(X) = (-1) * pq + 0 + (+1) * pq = 0$, which is obvious as the distribution is symmetric. $E(X^2) = pq + pq = 2pq$, from which the variance follows as $2pq - 0 = 2pq$. Note that this is not a binomial distribution as the probabilities are not obtained as the terms of a binomial expansion. This type of distribution finds applications in digital signal processing (DSP), cryptography, and many other fields.

Consider a Gaussian noise generator. If samples are taken and converted to bits using the rule $b(i) = +1$ if $g(i) \geq \theta$ and $b(i) = -1$, otherwise, where θ is a threshold, and $g()$ denotes a sampled signal. Now associate a probability p to the first event and complementary probability to the other event. If we add n of the bits together modulo-2, the distribution of the sum can be modeled as a binomial distribution.

2.4.2   MOMENT RECURRENCES

Recurrence relations are expressions involving successive terms of a sequence. This can be used to evaluate any desired term using known low-order terms. Low-order moments of $\mathrm{BINO}(n, p)$ could be obtained using the recurrences

$$f_x(n, p)/f_x(n - 1, p) = (n/(n - x))q, \tag{2.8}$$

which upon cross multiplication and rearrangement becomes

$$x f_x(n, p) = n[f_x(n, p) - q f_x(n - 1, p)]. \tag{2.9}$$

Multiply both sides by x^k, denote the moments by $\mu_k(n, p)$, and sum over the range of x to get

$$\mu'_{k+1}(n, p) = n \left[\mu'_k(n, p) - q * \mu'_k(n - 1, p) \right] \quad \text{with} \quad \mu'_0(n, p) = 1. \tag{2.10}$$

Put $k = 1$, to get $\mu'_2(n, p) = n[\mu'_1(n, p) - q\mu'_1(n - 1, p)]$. Substituting $\mu'_1(n, p) = np$, $\mu'_1(n - 1, p) = (n - 1)p$, the RHS becomes $n[np - q(n - 1)p] = np[np + q]$. Higher-order moments are obtained similarly.

Example 2.9   Binomial ordinary moment recurrence   Prove that the ordinary moments of a binomial distribution satisfy the recurrence

$$\mu'_{k+1} = p \left(q \frac{\partial}{\partial p} \mu'_k + n \mu'_k \right). \tag{2.11}$$

Solution: Write

$$\mu'_k = \sum_{x=0}^{n} x^k \binom{n}{x} p^x (1-p)^{n-x}. \tag{2.12}$$

Differentiate (2.12) by p to get

$$\frac{\partial}{\partial p}\mu'_k = \sum_{x=0}^{n} x^k \binom{n}{x} \left[(1-p)^{n-x} x p^{x-1} - p^x (n-x)(1-p)^{n-x-1}\right]. \tag{2.13}$$

Write $q = 1 - p$ and consider the terms $q^{n-x} x p^{x-1} + x p^x q^{n-x-1} = x p^{x-1} q^{n-x-1}[q + p] = x p^{x-1} q^{n-x-1}$. Substitute in (2.13) and simplify to get

$$\frac{\partial}{\partial p}\mu'_k = \sum_{x=0}^{n} x^{k+1} \binom{n}{x} \left[q^{n-x-1} p^{x-1} - n \sum_{x=0}^{n} x^k \binom{n}{x} p^x q^{n-x-1}\right]$$

$$= \mu'_{k+1}/(pq) - (n/q)\mu'_k. \tag{2.14}$$

Multiply throughout by pq and rearrange to get $\mu'_{k+1} = p\left(n\mu'_k + q\frac{\partial}{\partial p}\mu'_k\right)$.

Example 2.10 Binomial central moment recurrence   Prove that the central moments of a binomial distribution satisfy the recurrence

$$\mu_{k+1} = pq\left(\frac{\partial}{\partial p}\mu_k + nk\mu_{k-1}\right). \tag{2.15}$$

Solution: Consider

$$\mu_k = \sum_{x=0}^{n} (x - np)^k \binom{n}{x} p^x q^{n-x}. \tag{2.16}$$

Note that there are three functions of p on the RHS (as $q = 1 - p$). Differentiate both sides w.r.t. p using chain rule to get

$$\frac{\partial}{\partial p}\mu_k = -nk \sum_{x=0}^{n} (x-np)^{k-1} \binom{n}{x} p^x q^{n-x} + \sum_{x=0}^{n} (x-np)^k \binom{n}{x} x p^{x-1} q^{n-x}$$

$$- \sum_{x=0}^{n} (x-np)^k \binom{n}{x} (n-x) p^x q^{n-x-1}$$

$$= -nk\mu_{k-1} + \sum_{x=0}^{n} (x-np)^k \binom{n}{x} p^{x-1} q^{n-x-1} (xq - (n-x)p). \tag{2.17}$$

Write $xq - (n - x)p$ as $x(1 - p) - (n - x)p = x - np$ and combine with $(x - np)^k$; multiply both numerator and denominator by pq to get $-nk\mu_{k-1} + \mu_{k+1}/pq$. Rearrange the expression to get the result.

Example 2.11   Binomial central moment recurrence   Prove that the central moments of a binomial distribution satisfy the recurrence

$$\mu_{k+1}(n, p) = nq \left[\mu_k(n, p) - \sum_{j=0}^{k} \binom{k}{j} (-p)^{k-j} \mu_j(n - 1, p) \right]. \tag{2.18}$$

Solution: Write

$$\mu_k = \sum_{x=0}^{n} (x - np)^k \binom{n}{x} p^x (1 - p)^{n-x}. \tag{2.19}$$

Consider Equation (2.9) as $(n - x) f_x(n, p) = nq f_x(n - 1, p)$. Write $(n - x)$ as $n(p + q) - x = nq - (x - np)$ on the LHS. Multiply throughout by $(x - np)^k$, write $(x - np) = [(x - (n - 1)p) - p]$ on the RHS, expand using binomial theorem and sum over the proper range to get

$$nq \, \mu_k(n, p) - \mu_{k+1}(n, p) = nq \sum_{x} \sum_{j=0}^{k} \binom{k}{j} (x - (n - 1)p)^j (-p)^{k-j} f_x(n - 1, p)$$

$$= nq \sum_{j=0}^{k} \binom{k}{j} (-p)^{k-j} \mu_j(n - 1, p). \tag{2.20}$$

Rearrange (2.20) to get

$$\mu_{k+1}(n, p) = nq \left[\mu_k(n, p) - \sum_{j=0}^{k} \binom{k}{j} (-p)^{k-j} \mu_j(n - 1, p) \right] \tag{2.21}$$

with $\mu_0(n - 1, p) = 1$.

Theorem 2.12   *The factorial moment $\mu_{(r)}$ is given by $\mu_{(r)} = n^{(r)} p^r$.*

Proof.   Consider the expression for factorial moment as

$$\mu_{(r)} = E[x_{(r)}] = E[x(x - 1) \ldots (x - r + 1)] = \sum_{x} [x(x - 1) \ldots (x - r + 1)] f(x). \tag{2.22}$$

Substitute the PMF and sum over the proper range of x to get the RHS as

$$\sum_x x(x-1)\dots(x-r+1)n!/(x!(n-x)!)p^x q^{(n-x)}. \tag{2.23}$$

Write $n!$ in the numerator as $n(n-1)\dots(n-r+1)*(n-r)!$, this becomes

$$n(n-1)\dots(n-r+1)p^r\sum_x(n-r)!/((x-r)!(n-x)!)p^{x-r}q^{(n-x)} = n^{(r)}p^r. \tag{2.24}$$

Write

$$x(x-1)\dots(x-r+1)/(n-x)! = \sum_{i=0}^{r}s(r,i)x^i, \tag{2.25}$$

where $s(r,i)$ is the Stirling number of first kind. The factorial moments are found using Stirling numbers as

$$\mu_{(r)} = \sum_{i=0}^{r}s(r,i)\mu_i', \tag{2.26}$$

where μ_i' denotes the ith ordinary moment. The reverse relationship is

$$x^r = \sum_{i=0}^{r}S(r,i)x!/(x-i)!, \tag{2.27}$$

where $S(r,i)$ is the Stirling number of second kind. This allows us to write

$$\mu_r' = \sum_{k=0}^{r}S(r,k)\mu_{(k)}'. \tag{2.28}$$

As the domain of the random variable includes zero, the first negative moment is defined as $E(1/(1+X))$.   □

Example 2.13   Inverse moment of Binomial distribution   Find the first inverse moment of a binomial distribution.

Solution: By definition

$$E(1/(X+1)) = \sum_{x=0}^{n}(1/(x+1))\binom{n}{x}p^x q^{n-x}. \tag{2.29}$$

Expand $\binom{n}{x}$ and write this as

$$E(1/(X+1)) = \sum_{x=0}^{n}n!/[(x+1)\,x!(n-x)!]p^x q^{n-x}. \tag{2.30}$$

Multiply and divide the RHS by $(n + 1)p$ and write it as

$$E(1/(X + 1)) = (1/[(n + 1)p]) \sum_{x=0}^{n} (n + 1)!/[(x + 1)!(n - x)!] p^{x+1} q^{n-x}. \tag{2.31}$$

The sum on the RHS is $(p + q)^{n+1}$ less the first term $\binom{n+1}{0} p^0 q^{n+1-0} = q^{n+1}$. As $(p + q) = 1$, the RHS becomes $(1 - q^{n+1})/[(n + 1)p]$. This is the first inverse moment of BINO(n, p).

Problem 2.14   If $X \sim$ BINO(n, p), find $E(1/(X + 1)(X + 2))$. Hint: Proceed as above, multiply and divide RHS by $(n + 1)(n + 2)p^2$, and identify missing terms in the beginning of $(p + q)^{n+2}$.

Theorem 2.15   *Prove that the mean deviation from the mean of a binomial distribution is*

$$2npq \binom{n - 1}{v} p^v q^{n-v-1}, \tag{2.32}$$

where v is the largest integer less than np (symbolically $v = \lfloor np \rfloor$).

Proof. By definition,

$$\mathrm{MD} = \sum_{x=0}^{n} |x - np| \binom{n}{x} p^x q^{n-x}. \tag{2.33}$$

Split the RHS into two sums for $x \le v$ and $x > v$, respectively:

$$\sum_{x=0}^{v} (np - x) \binom{n}{x} p^x q^{n-x} + \sum_{x=v+1}^{n} (x - np) \binom{n}{x} p^x q^{n-x}. \tag{2.34}$$

As $E(X - \mu) = 0$, we have $\sum_{x=0}^{n} (np - x) \binom{n}{x} p^x q^{n-x} = 0$. Substitute in the above to get

$$2 \sum_{x=0}^{v} (np - x) \binom{n}{x} p^x q^{n-x}. \tag{2.35}$$

Write $(np - x) = (np - x(p + q)) = p(n - x) - xq$, and split the above sum as

$$2 \left(p \sum_{x=0}^{v} (n - x) \binom{n}{x} p^x q^{n-x} - q \sum_{x=0}^{v} x \binom{n}{x} p^x q^{n-x} \right). \tag{2.36}$$

Expand $\binom{n}{x}$ and cancel out common terms. Take nq outside the summation to get

$$\mathrm{MD} = 2 \left(npq \sum_{x=0}^{v} \binom{n - 1}{x} p^x q^{n-x-1} - npq \sum_{x=1}^{v} \binom{n - 1}{x - 1} p^{x-1} q^{n-x} \right). \tag{2.37}$$

Taking npq as common factor, we notice that the alternate terms in the left and right sums of (2.30) cancel out giving

$$\text{MD} = 2npq \binom{n-1}{v} p^v q^{n-v-1}, \tag{2.38}$$

where $v = \lfloor np \rfloor$ is an integer. This is the form obtained originally by De Moivre (1730) and subsequently by Bertrand (1889). As the variance of the binomial distribution is npq, the ratio MD/σ^2 is $2\binom{n-1}{v} p^v q^{n-v-1}$, which is twice the PMF of $\text{BINO}(n-1, p)$ at $x = v$. If np is non-integer, we could use the gamma function to get

$$\text{MD} = 2(\Gamma(n+1)/[v!\,\Gamma(n-v)])\,p^{v+1} q^{n-v}, \tag{2.39}$$

where $q = 1 - p$. Johnson [37], among others, have discussed other equivalent forms and approximations. As np appears as a multiplier in (2.38), the results are accurate without a correction factor mentioned in Chapter 1. This reduces to $n\binom{n-1}{v}/2^n$ for $p = 1/2$. □

2.4.3   ADDITIVITY PROPERTY

Theorem 2.16   *If $X_1 \sim \text{BINO}(n_1, p)$ and $X_2 \sim \text{BINO}(n_2, p)$ are independent binomial random variables with the same probability of success p, the sum $X = X_1 + X_2$ is distributed as $\text{BINO}(n_1 + n_2, p)$.*

Proof. The easiest way to prove the above result is using the moment generating function (MGF). As X_1 and X_2 are independent, $M_{X_1+X_2}(t) = M_{X_1}(t) * M_{X_2}(t)$. Substituting $M_X(t) = (q + pe^t)^n$, the RHS becomes $(q + pe^t)^{n_1} * (q + pe^t)^{n_2} = (q + pe^t)^{n_1+n_2}$, which is the MGF of $\text{BINO}(n_1 + n_2, p)$. An interpretation of this result in terms of Bernoulli trials is the following—"if there are n_1 independent Bernoulli trials with the same probability of success p, and another n_2 independent Bernoulli trials with the same probability of success, they can be combined in any desired order to produce a binomial distribution of size $n_1 + n_2$."

Another way of stating the above theorem is that if $X + Y$ is distributed as $\text{BINO}(n_1 + n_2, p)$, and either of X or Y is distributed as $\text{BINO}(n_1, p)$, the other random variable must be $\text{BINO}(n_2, p)$ (or $\text{BER}(p)$ is $n_2 = 1$). This result can be extended to any number of independent binomial distributions with the same probability of success p. Symbolically, if $X_i \sim \text{BINO}(n_i, p)$, then $S = \sum_i X_i \sim \text{BINO}(\sum_i n_i, p)$ (Table 6.4). □

Table 2.3: Properties of binomial distribution

Property	Expression	Comments
Range of X	$x = 0, 1,\ldots,n$	Discrete, finite
Mean	$\mu = np$	Need not be integer
Variance	$\sigma^2 = npq = \mu q$	$\Rightarrow \mu > \sigma^2$
CV	$(q/np)^{1/2}$	$\sqrt{q/(np)}$
Mode	$(x - 1), x$ if $(n + 1)*p$ is integer	$x = (n + 1)*p$ else
Skewness	$\gamma_1 = (1 - 2p)/\sqrt{npq}$	$= (q - p)/\sqrt{npq}$
Kurtosis	$\beta_2 = 3 + (1 - 6pq)/npq$	
MD	$2npq(\binom{n-1}{\lfloor np \rfloor}p^{\lfloor np \rfloor}q^{n-1-\lfloor np \rfloor})$	$2\sum_{x=0}^{\lfloor np-1 \rfloor} I_{1-p}(n - x, x + 1)$
$E[X(X-1)\ldots(X-r+1)]$	$n^{(r)}p^r$	FMGF $= (1 + pt)^n$
MGF	$(q + pe^t)^n$	$= (1/2)^n (1 + e^t)^n$ if $p = q$
PGF	$(q + pt)^n$	$= ((1 + t)/2)^n$ if $p = q$
Additivity	$\sum_{i-1}^m B(n_i, p) = B(\sum_{i-1}^m n_i, p)$	Independent
Recurrence	$f(x;n,p)/f(x-1;n,p)=(n-x+1)p/[xq]$	$1 + ((n + 1)p - x)/(xq)$
Tail probability	$\sum_{x=k}^n \binom{n}{x}p^x q^{n-x}$	$I_p(k, n - k + 1)$
	Symmetric when $p = q = 1/2$.	

2.4.4   MD OF BINOMIAL DISTRIBUTION

The MD of binomial distribution can be found using the incomplete beta function. We know that the mean of binomial distribution is np. Hence, the MD is given by the power method in page 11 as

$$\text{MD} = 2 \sum_{x=ll}^{\lfloor np \rfloor} F(x) = 2 \sum_{x=0}^{\lfloor np \rfloor} I_{1-p}(n - x, x + 1). \tag{2.40}$$

The results obtained by (2.40) and Theorem 1.7 are given in the Table 2.3 (see also Table 2.1). Both results totally tally when np is an integer or half-integer (with the correction term). Otherwise we have to add a correction term $2\delta F(\lfloor \mu \rfloor)$ where δ is the fractional part. If n is large and p is small, the first few $F(x)$ terms in (2.40) could be nearly zeros. A solution is to start the iterations at $x = \lfloor np \rfloor$ and recur backward, or to start the iterations at a higher integer from lower limit ll and recur forward.

Table 2.4: Mean deviation of binomial distribution using power method

n, p	0	1	2	3	4	5	6	7	Eq. (2.40)
10, 0.10	0.349								0.6974
15, 0.40	0.000	0.005	0.027	0.091	0.217	0.4032			1.4875
20, 0.50	0.000	0.001	0.006	0.021	0.058	0.132	0.252	0.412	1.762
30, 0.10	0.042	0.184	.4114						1.2749
40, 0.05	0.129	0.399							1.0552
50, 0.14	0.001	0.005	0.022	0.067	0.153	0.281	0.438		1.934
8, 0.875	0.000	0.000	0.000	0.001	0.011	0.067	0.264		0.6872
50, 0.20	0.001	0.006	0.018	0.048	0.103	0.190	0.307	0.444	2.2371
80, .1125	0.001	0.004	0.016	0.046	0.102	0.191	0.310	0.448	2.2339
25, 0.36	0.000	0.002	0.007	0.025	0.068	0.148	0.271	0.425	1.8937

First column gives n, p values of binomial. Second column onward are the values computed using Equation (2.40). Last column finds the MD using equation Power method, which is the same as that found using Equation (2.40) when np is an integer. When np is not an integer, add the correction term mentioned above. Values have been left shifted by two places for (20, 0.50), (50, 0.20) rows, and by one place to the left for few other rows, as the first few entries are zeros.

2.4.5   NEW DISTRIBUTION FROM MD

Let $X \sim \text{BINO}(n, p)$. If np is an integer, the MD is given by $\text{MD} = 2 \sum_{x=ll}^{\lfloor np \rfloor} F(x)$, where $ll = 0$ is the lower limit. Define a new PMF $f_1(x; n, c, p)$ as

$$f_1(x; n, c, p) = (2/c) \sum_{k=0}^{x} \binom{n}{k} p^k (1-p)^{n-k}, \quad x = 0, 1, \ldots, \lfloor np \rfloor, \qquad (2.41)$$

where $c = $ MD of the original distribution. This is a well-defined distribution because $f_1(x; n, c, p) \geq 0$, and the sum over the range is one. If the MD of this new distribution is MD1, we could continue obtaining new distributions by repeating the above argument:

$$f_2(x; n, c, p) = (2/\text{MD1}) \sum_{k=0}^{x} f_1(k; n, c, p) \quad x = 0, 1, \ldots, \lfloor \mu_1 \rfloor, \qquad (2.42)$$

where μ_1 is the mean of the distribution with PMF $f_1(x; n, c, p)$. Thus, we could obtain many new distributions.

2.4.6   TESTS OF HYPOTHESIS

Consider tests of hypotheses on proportions. If the null hypothesis is composite, the level of significance (α) is the maximum probability of rejecting the null. If we take p as a variable $(0 < p \leq \alpha = .05)$ (say), $\Pr[\text{reject } H_0 \text{ when true}] = \sum_{j=k}^{n} \binom{n}{j} p^j q^{n-j}$ where k is an integer such that the number of defectives in a sample of size n is less than p. This is the SF of a binomial distribution. Another application of binomial distribution is in tests of hypotheses about quantiles. If a random sample is available for our test, we will first rearrange the sample values in increasing order and denote by $x_{(r)}$ the rth order statistic.

Similarly, $100(1 - \alpha)\%$ confidence intervals for binomial proportions are found by solving the relation

$$\sum_{x=k}^{n} \binom{n}{x} p_0^x (1 - p_0)^{n-x} = \alpha \tag{2.43}$$

which is again the SF of a binomial distribution $\text{BINO}(n, p_0)$. This can be solved either numerically, or analytically using the relationship between CDF of binomial distribution and incomplete beta or F distribution. As the binomial tail area is related to the F distribution, the above relation can be solved using F distribution tables:

$$k(1 - p_0)/[(n - k + 1)p_0] = F(\alpha/2, 2n - 2k + 2, 2k) \tag{2.44}$$

from which we get

$$p_0 = 1/[1 + (n - k + 1)F/k]. \tag{2.45}$$

2.4.7   BINOMIAL CONFIDENCE INTERVALS

Confidence intervals (CI) are range of values that an unknown parameter is likely to take with a fixed probability. Confidence intervals for proportions in research studies are often reported in terms of standard normal distribution as

$$\hat{p} \pm z_{1-\alpha/2}(pq/n)^{1/2}, \tag{2.46}$$

where $z_{1-\alpha/2}$ denotes percentile points of a standard normal distribution with cut-off $1 - \alpha/2$ area and $q = 1 - p, \alpha$ is the desired confidence, n is the sample size, and p the proportion of interest. These are developed under the assumption of approximate normality of the proportion, so that $z_{1-\alpha/2} = 1.96$ for 95% confidence, $z_{1-\alpha/2} = 2.58$ for 99% confidence. Exact CI can be obtained using binomial CDF as

$$\sum_{k=0}^{x} \binom{n}{k} p_0^k (1 - p_0)^{n-k} = \alpha/2 \sum_{k=x}^{n} \binom{n}{k} p_1^k (1 - p_1)^{n-k} = 1 - \alpha/2. \tag{2.47}$$

2.5   GENERATING FUNCTIONS

Generating functions are extensively discussed in [18]. They are useful tools to find probabilities, moments, cumulants, and mean deviations of discrete distributions.

2.5.1   PROBABILITY GENERATING FUNCTIONS

Theorem 2.17   *The PGF is $P_X(t) = (q + pt)^n$, and the MGF is $M_X(t) = (q + pe^t)^n$.*

Proof. The PGF is

$$E(t^x) = \sum_{x=0}^{n} t^x \binom{n}{x} p^x q^{n-x} = \sum_{x=0}^{n} \binom{n}{x} (pt)^x q^{n-x} = (q + pt)^n, \qquad (2.48)$$

where $E()$ is the expectation operator. The MGF and characteristic function (ChF) are found by replacing t^x by e^{tx} and e^{itx} to get $M_x(t) = (q + p\exp(t))^n$, and $\phi_x(t) = (q + p\exp(it))^n$. $\square$

2.5.2   VARIANCE GENERATING FUNCTION (VGF)

A GF can be developed for the variance of a BINO(n, p) distribution as follows. Consider $V(t; n, p) = pq/(1 - t)^2$. Expand as an infinite series to get

$$V(t; n, p) = pq \left(1 + 2t + 3t^2 + \cdots + nt^{n-1} + \cdots\right). \qquad (2.49)$$

This shows that the coefficient of t^n in $pqt/(1 - t)^2$ is npq (i.e., coefficient of t^{n-1} in $pq/(1 - t)^2$). Thus, the VGF of BINO(n, p) is $pqt/(1 - t)^2$.

Example 2.18   Identical defect proportion   Two machines M1 and M2 produce ball-bearings with the same defect rate of 0.01, but M1 can produce 10,000 and M2 can produce 12,000 almost identically looking ball bearings at full capacity. A quality control inspector randomly selects 7 bearings produced by M1 and 8 bearings produced by M2. What is the probability of the following: (i) exactly one defective, and (ii) at least two defectives?

Solution: As the probability of defectives are the same in both machines, we combine the samples and assume $n = 7 + 8 = 15$. Probability of exactly one defective $= \binom{15}{1}(0.01)^1(0.99)^{14} = 0.13031187$. In the second part, we find the complementary probability and subtract from one to get $1 - \Pr[\text{no defectives}] - \Pr[\text{one defective}] = 1 - (0.99)^{15} - 0.13031187 = 0.00962977$.

Example 2.19   Dual binomial distribution   If $X \sim \text{BINO}(n, p)$, derive the distribution of $Y = n - X$, and obtain its PGF and MGF. Obtain the mean and variance. What is the additive property for Y?

Solution: As X takes the values $0, 1, \ldots, n$. Y also takes the same values in reverse. Thus, the range of X and Y are the same:

$$\Pr[Y = y] = P[X = n - y] = \binom{n}{n-y} p^{n-y} q^{n-(n-y)} = \binom{n}{y} q^y p^{n-y} \qquad (2.50)$$

using $\binom{n}{n-y} = \binom{n}{y}$.

This is the PMF of a binomial distribution with p and q reversed. It is called dual binomial distribution. Hence, $Y = n - X \sim \mathrm{BINO}(n, q)$, so that all properties are obtained by swapping the roles of p and q in the corresponding property of $\mathrm{BINO}(n, p)$. In particular, sum of the left-tail probabilities (CDF) of $\mathrm{BINO}(n, p)$ is the sum of right-tail probabilities (SF) of $\mathrm{BINO}(n, q)$. In other words, $\Pr[X \leq k] = \Pr[Y \geq n - k]$ where X is binomial and Y is dual binomial distributions (this result is used in Chapter 5). The PGF is

$$P_Y(t) = E(t^y) = \sum_{y=0}^{n} t^y \binom{n}{y} q^y p^{n-y} = \sum_{y=0}^{n} \binom{n}{y} (qt)^y p^{n-y} = (qt + p)^n = (p + qt)^n.$$

Similarly, the MGF is $M_Y(t) = E(e^{ty}) = (p + qe^t)^n$. From this the cumulant generating function (CGF) follows as $n \ln(p + qe^t)$. The mean and variance are nq and npq, respectively. This shows that X and Y have the same variance, but the mean is $nq = n(1 - p) = n - np$. As it is binomial distributed, the additive property remains the same. Thus if Y and Z are $\mathrm{BINO}(n_1, q)$ and $\mathrm{BINO}(n_2, q)$ random variates, then $Y + Z$ is distributed as $\mathrm{BINO}(n_1 + n_2, q)$, provided q (or equivalently $p = 1 - q$) is the same. Thus, the binomial distribution is decomposable into a finite number of identical distributions. The CDF of X and Y are related as $F(y) = \Pr[Y \leq y] = \Pr[n - X \leq y] = \Pr[X \geq n - y] = 1 - \Pr[X < n - y] = 1 - F_x(n - y - 1; n, p)$. These are summarized in Table 2.3.

Problem 2.20   If $X \sim \mathrm{BINO}(n, p)$ and $Y = n - X$, find $E(XY)$.

2.6   ALGORITHM FOR COMPUTING BINOMIAL DISTRIBUTION

Successive probabilities of the binomial distribution are found using the recurrence relationship $f(x)/f(x - 1) = ((n - x + 1)/x) p/q$ with starting value $f(0) = \binom{n}{0} q^n = q^n$. This could also be written as

$$f(x)/f(x - 1) = [(n + 1)/x - 1] \, p/q \quad \text{or as} \quad 1 + [(n + 1)p - x]/(qx), \qquad (2.51)$$

where the last expression is obtained by adding and subtracting $1/q$, writing $-1/q$ as $-x/qx$, and using $(-p/q + 1/q) = 1$. When n is very large and $x > n/2$, we could reduce the number of iterations by starting with $f(n) = \binom{n}{n} p^n = p^n$, and recurring backward using the relationship $f(x-1) = \frac{q}{p} \frac{x}{(n-x+1)} f(x)$. These forward recurrences are useful when several probabilities are to be evaluated simultaneously. Backward recurrences could also be developed to compute the probabilities in reverse using

$$f(x-1) = [x/(n-x+1)](q/p)f(x). \tag{2.52}$$

Example 2.21   Defective component   One in 200 manufactured electronic components is known to be defective. If a sample of 12 components are chosen, find the probability of finding zero, one, and two defectives.

Solution: Several probabilities are to be evaluated in this problem. Thus, we resort to the recurrence formula. Here $p = 1/200$ and $q = 199/200$, so that $p/q = 1/199$. Then, $\Pr[\text{no defective}] = p_0 = q^n = (199/200)^{12} = 0.9416228$. From this $\Pr[\text{one defective}] = p_1$ is found as $p_1 = p_0 * ((n-x+1)/x)p/q$ where $x = 1$ to get $p_1 = 0.9416228 * 12 * 1/199 = 0.056781$. Next, $p_2 = p_1 * ((n-x+1)/x)p/q$, for $x = 2$ gives $0.056781 * (11/2) * (1/199) = 0.0015693$.

Example 2.22   Magnetic field   Consider a magnetic field F in which there are n magnets placed. Each magnet undergoes half-spin or full-spin, so that each magnetic moment points upward (parallel to F) or downward (opposite direction). If p is the probability that any magnet points upward independent of others, what is the probability that exactly half of them point upwards (assume that n is even).

Solution: As the magnets are acting independently, the desired probability is given by the binomial distribution for $x = n/2$ as $\binom{n}{n/2} p^{n/2} (1-p)^{n/2} = \binom{n}{n/2} (pq)^{n/2}$ where $q = 1 - p$. When the field is zero ($F = 0$), $p = q = 1/2$ so that this becomes $\binom{n}{n/2}(1/4)^{n/2} = \binom{n}{n/2}(1/2)^n$.

Example 2.23   Winning group   A class has b boys and g girls, both ≥ 2. A competition is conducted between the boys (who form group G1) and girls (who form group G2), where each game is independent of others and it is between the groups. If there are n prizes to be distributed to the winning groups, find (i) probability that girls bag more prizes than boys, (ii) number of prizes bagged by boys is odd, (iii) number of prizes bagged by girls is even number, and (iv) boys get no prizes.

Solution: As there are b boys and g girls, the proportion of boys is $b/(b+g)$ and that of girls is $g/(b+g)$. As there are n prizes, the distribution of the prizes in favor of the boys is a $\text{BINO}(n, b/(b+g))$, where we have assumed that a "success" corresponds to the boys winning a prize. We assume that this probability remains the same because the prizes are distributed to the groups independently.

(i) Probability that girls bag more prizes than boys = Probability that boys get less prizes than girls $= \Pr[x < n - x] = \Pr[2x < n] = \Pr[x < n/2] = \sum_{i=0}^{\lfloor n/2 \rfloor} \binom{n}{i} p^i q^{n-i}$, where $p = b/(b + g)$, and the summation is from 0 to $(n-1)/2$ if n is odd, and to $n/2$ if n is even. (ii) The number of prizes bagged by boys is odd $= \Pr[x = 1, 3, \ldots] = \sum_{x \, \text{odd}} \binom{n}{x} p^x q^{n-x}$. To evaluate this sum, consider the expression $(p + q)^n - (p - q)^n$. Expanding using binomial theorem and canceling out all even terms we get

$$(p + q)^n - (p - q)^n = 2 \left[\sum_{x \, \text{odd}} \binom{n}{x} p^x q^{n-x} \right]. \qquad (2.53)$$

Hence, the required probability is $\frac{1}{2}[(p + q)^n - (p - q)^n]$. But $p + q = 1$, and $p - q = b/(b + g) - g/(b + g) = (b - g)/(b + g)$. Substitute in the above to get the required probability as $\frac{1}{2}[1 - [(b - g)/(b + g)]^n]$. When the number of boys is less than that of girls, the second term can be negative for odd n. (iii) Number of prizes bagged by girls is even $= \left[\sum_{x \, \text{even}} \binom{n}{x} q^x p^{n-x} \right]$ where we have swapped the roles of p and q. To evaluate this sum, consider the expression $(p + q)^n + (q - p)^n$. Expanding using binomial theorem, all odd terms cancel out giving $2 \left[\sum_{x \, \text{even}} \binom{n}{x} q^x p^{n-x} \right]$. Hence, the required probability is $\frac{1}{2}[1 + [(g - b)/(b + g)]^n]$. (iv) Probability that boys get no prizes $= q^n = [g/(b + g)]^n$.

2.6.1 TAIL PROBABILITIES

The CDF of a binomial distribution BINO(n, p) is $F_X(n, p) = \sum_{k=0}^x \binom{n}{k} p^k q^{(n-k)}$. We could compute this by the straightforward method of adding the successive probabilities. However for large n and k, this method is very inefficient. A better approach is to use the relationship between the binomial distribution and the incomplete beta function as follows:

$$F_X(k; n, p) = P[X \leq k] = \sum_{x=0}^k \binom{n}{x} p^x q^{n-x} = I_{1-p}(n - k, k + 1). \qquad (2.54)$$

The LHS of (2.54) is a discrete sum whereas the RHS is a continuous function of p. This can also be expressed in terms of F distribution as

$$F_X(k; n, p) = P[X \leq k] = F_{1-p(k+1)/[p(n-k)]}(2(n - k), 2(k + 1)), \qquad (2.55)$$

where both degrees of freedom parameters are even. When $k > np/2$, this is computed as

$$F_X(k) = 1 - I_p(k + 1, n - k). \qquad (2.56)$$

Alternatively, the tail probabilities (SF) can be expressed as

$$\sum_{x=k}^n \binom{n}{x} p^x q^{n-x} = \frac{n!}{(k - 1)!(n - k)!} \int_{y=0}^p y^{k-1}(1 - y)^{n-k} dy. \qquad (2.57)$$

Replacing the factorials by gamma functions, this is seen to be equivalent to

$$\sum_{x=k}^{n} \binom{n}{x} p^x q^{n-x} = I_p(k, n-k+1). \tag{2.58}$$

The discrepancy between the above two formulas is due to the fact that $\Pr[x = k]$ is included in the SF while it is omitted in the other. As the incomplete beta function is widely tabulated, it is far easier to evaluate the RHS. This is especially useful when n is large and k is not near n.

Suppose the mean $\overline{X} = \sum_i X_i/n$ is considered as a new random variable. The possible values assumed are $0, 1/n, 2/n, \ldots (n-1)/n$. Then $\Pr[\overline{X} \leq k] = \Pr[S_n \leq nk]$. As the sum has a binomial distribution, this can be evaluated as the CDF of a BINO(n, p) up to $\lfloor nk \rfloor$. The normal approximation can be used to evaluate binomial tail probabilities in two situations: (i) when p is close to 0.5 and n is reasonably large (say greater than 10), and (ii) when $n \to \infty$ and p is not very small (say np and nq are both ≥ 15). Symbolically, the random variable $Z = (X - np)/(npq)^{1/2}$ is approximately standard normal distributed as $n \to \infty$. This can be expressed as

$$\sum_{x=a}^{b} \binom{n}{x} p^x (1-p)^{n-x} \simeq \int_{c}^{d} (1/\sqrt{2\pi}) \exp(-z^2/2) dz = \Phi(d) - \Phi(c), \tag{2.59}$$

where $c = (a - np)/\sqrt{np(1-p)}$, and $d = (b - np)/\sqrt{np(1-p)}$, $\Phi()$ denotes the CDF of standard normal distribution, and $a < b$ are integers. Note that this approximation is good only when either n is large or p is very close to $1/2$. It can also be used to approximate individual probabilities as

$$\binom{n}{x} p^x q^{n-x} \simeq 1/(2\pi npq)^{1/2} \exp(-(x - np)^2/(2npq)). \tag{2.60}$$

The proportion of success (e.g., proportion of defective items) can also be approximated as $Z = (X/n - p)/(pq/n)^{1/2} \sim N(0, 1)$.

A bound for undetected errors may be desired in some data transmission and archival systems like image archival. Although built-in CRC (Cyclic Redundancy Check) algorithms can be helpful to detect and correct errors to an extent, the image quality may be acceptable up to some threshold. The following example derives a bound when n bits are transmitted and x denotes the number of bits received incorrectly (so that p = probability of erroneous data transmission).

Example 2.24 Bounds on SF of binomial distribution Prove that the binomial survival function is upper-bounded by $((x+1)q/[(x+1) - (n+1)p] f(x; n, p)$.

Solution: By definition SF $S(x) = \sum_{x=k}^{n} \binom{n}{x} p^x q^{n-x}$ (we sum from k onward instead of $k +$ 1). Write this in expanded form as

$$\binom{n}{k} p^k q^{n-k} + \binom{n}{k+1} p^{k+1} q^{n-k-1} + \cdots + \binom{n}{n} p^n. \tag{2.61}$$

Take $\binom{n}{k} p^k q^{n-k}$ as a common factor and write $\binom{n}{k+1} = (n-k)/k \binom{n}{k}$. Although the above is a finite sum, we will assume that it goes to infinity as we are seeking an upper bound. This gives $S(x) < \binom{n}{k} p^k q^{n-k} [1 - (n-k)p/(kq)]^{-1}$.

Problem 2.25   Prove that the median m of BINO(n, p) satisfies $|m - np| \leq \max(p, 1 - p)$.

Example 2.26 Online quiz   An online quiz comprises of 15 multiple-choice questions, each with 5 answer choices A, B, C, D, E. Assume that students are allowed to take it anywhere using their smartphone. Find the probability for each of the following events if the pass minimum is eight correct answers: (i) a student guesses all the questions, and (ii) a student knows four correct answers and guesses all the rest of the questions.

Solution: As there are five answer choices for each question, the probability that a student who guesses get the correct answer is $p = 1/5$. Then the probability that a student gets eight or more correct answers is $\sum_{k=8}^{15} \binom{15}{k}(1/5)^k (4/5)^{15-k}$. In part (ii) the student knows correct answer to 4 questions so that 11 questions need guessing. As the pass mark is eight, the student need to get $8 - 4 = 4$ or more marks to pass. The probability of this is $\sum_{k=4}^{11} \binom{11}{k}(1/5)^k (4/5)^{11-k}$.

Example 2.27 Quiz with penalty   Redo the above problem if each incorrect guess carries a penalty of 1/4 points.

Solution: Assume that the student guesses all questions. If 9 questions are guessed correctly, the wrong 6 questions will result in a negative score of $6/4 = 1.5$ so that the net marks is $9.0 - 1.5 = 7.5$. Thus, the minimum number of correct answers required to pass is 10 (because $10 - 5/4 = 8.75 > 8$). The required probability to pass is $\sum_{k=10}^{15} \binom{15}{k}(1/5)^k (4/5)^{15-k}$.

2.6.2   APPROXIMATIONS

As there are two parameters for this distribution, the binomial tables are lengthy and cumbersome. When the probability of success p is very close to .5, the distribution is nearly symmetric. The normal approximation is not good for x values away from the modal value (10) when n is small due to the disparity in the variance. If we reduce the variance of the approximating normal, the peak probabilities will increase. When n is large, the central limit theorem can be used to approximate binomial by a normal curve. The accuracy depends both upon n and whether the value of p is close to 0.5. Not only the probabilities, but the cumulative probabilities can also be approximated using normal tail areas. This approximation is quite good when p is near .5 rather than near 0 or 1 (use the normal approximation when $np > 10$ or $np(1-p) > 10$ or both np and $n(1-p)$ are > 5). Symbolically, $P[x \leq k] = Z(\frac{k-np}{\sqrt{npq}})$, where $Z()$ is the standard normal distribution. As this is an approximation of a discrete distribution by a continuous one, a continuity correction could improve the precision for small n. This gives us

$$P[x \leq k] = Z((k-np+.5)/\sqrt{npq}) \quad \text{and} \quad P[x \geq k] = 1 - Z((k-np-.5)/\sqrt{npq}).$$
(2.62)

2.7   LIMITING FORM OF BINOMIAL DISTRIBUTION

The binomial distribution tends to the Poisson law when $n \to \infty$, $p \to 0$ such that np remains a constant. This result was known to S.D. Poisson (1837), which is why the limiting distribution is called *Poisson distribution*. This is easily derived from the PMF as follows. Write the PMF as

$$f_X(k;n,p) = \binom{n}{k} p^k q^{n-k} = \frac{n(n-1)(n-2)\dots(n-k+1)}{k!} p^k q^{n-k}.$$
(2.63)

Multiply the numerator and denominator by n^k, combine it in the numerator with p^k, and write q^{n-k} as $(1-p)^n * (1-p)^{-k}$ to obtain:

$$f_X(k;n,p) = \frac{n}{n} \cdot \frac{n-1}{n} \cdot \frac{n-2}{n} \cdots \frac{n-k+1}{n} \frac{(np)^k}{k!} (1-p)^{-k}(1-p)^n.$$
(2.64)

According to our assumption, np is a constant (say λ) so that $p = \lambda/n$. Substitute in the above and let $n \to \infty$

$$\underset{n \to \infty}{Lt} f_X(k;n,p) = \underset{n \to \infty}{Lt} \frac{n}{n} \cdot \frac{n-1}{n} \cdot \frac{n-2}{n} \cdots \frac{n-k+1}{n} \frac{\lambda^k}{k!} (1-\lambda/n)^{-k}(1-\lambda/n)^n.$$
(2.65)

If k is finite, the multipliers all tend to 1, and $(1-\lambda/n)^{-k}$ also tends to 1. The last term tends to $e^{-\lambda}$ using the result that $\underset{n \to \infty}{Lt}(1-x/n)^n = e^{-x}$. Hence, in the limit, the RHS tends to the Poisson distribution $e^{-\lambda}\lambda^k/k!$. We could write this in more meaningful form as

$$f_X(x;n,p) = \binom{n}{x} p^x q^{n-x} \to e^{-np}(np)^x/x! + O(np^2/2) \quad \text{as} \quad n \to \infty,$$
(2.66)

where $O(np^2/2)$ is the asymptotic notation ([15], [48]). An interpretation of this result is that the binomial distribution tends to the Poisson law when $p \to 0$ faster than $n \to \infty$. In other words, the convergence rate is *quadratic* in p and *linear* in n. This allows us to approximate the binomial probabilities by the Poisson probabilities even for very small n values (say $n < 10$), provided that p is comparatively small.

The above result can also be proved using the PGF. We know that $P_x(t; n, p) = (q + pt)^n$. Write $q = 1 - p$, and take logarithm of both sides to get $\log(P_x(t; n, p)) = n \log(1 - p(1 - t))$. Write $n = -(-n)$ on the RHS, and expand as an infinite series using $-\log(1 - x) = x + x^2/2 + x^3/3 + \cdots$ to get

$$\log(P_x(t; n, p)) = -n\left[p(1 - t) + p^2(1 - t)^2/2 + p^3(1 - t)^3/3 + \cdots\right]. \tag{2.67}$$

Write $np = \lambda$, and take negative sign inside the bracket. Then the RHS becomes $\log(P_x(t; n, p)) = \lambda(t - 1) - np^2(t - 1)^2/2 + \cdots$. When exponentiated, the first term becomes the PGF of a Poisson distribution (page 111). The rest of the terms contain higher order powers of the form np^r/r for $r \geq 2$.

We have assumed that $p \to 0$ in the above proof. This limiting behavior of p is used only to fix the Poisson parameter. This has the implication that we could approximate both the left tail and right tail areas, as well as individual probabilities using the above approximation. If p is not quite small, we use the random variable $Y = n - X$, which was shown to have a binomial distribution (see example in page 37) with probability of success $q = 1 - p$, so that we could still approximate probabilities in both tails when p is very close to 1.

When p is near .5, a normal approximation is better than the Poisson approximation due to the symmetry. But a correction to the Poisson probabilities could improve the precision. For large values of n, the distributions of the count X and the sample proportion are approximately normal. This result follows from the central limit theorem. The mean and variance for the approximately normal distribution of X are np and $np(1 - p)$, identical to the mean and variance of the binomial(n, p) distribution. Similarly, the mean and variance for the approximately normal distribution of the sample proportion are p and $(p(1 - p)/n)$. The hypergeometric distribution asymptotically becomes the binomial distribution when both the parameters increase to infinity in such a way that $N/(N + M)$ remains a constant p. A proportionality relationship also exists among binomial distribution and beta distribution of first kind. Let $c = x + 1$ and $d = n - x + 1$. Then the beta and binomial distributions are related as $\text{BETA}(p; c, d) = (n + 1) \text{BINO}(x; n, p)$.

Example 2.28 Political parties   Consider a group of n individuals who support one of two political parties say P1 and P2. Assuming that none of the votes are invalid, what is the probability that a candidate of party P1 wins over the other candidate?

Solution: If the voting decision of an individual is not influenced by the decision of another (for example husband's and wife's decision or decision among friends) the proportion of individuals who support one of two political parties can be regarded as a binomial distributed random variable with probability p. To find winning chances, we need to consider whether n is odd or even. If n is odd, P1 will win if the number of votes received is $\geq \frac{n+1}{2}$. Thus, the required probability is $\sum_{x=\frac{n+1}{2}}^{n}\binom{n}{x}p^{x}q^{n-x}$, where p =probability that the vote is in favor of the candidate of P1. If n is even, the summation need to be carried out from $(n/2)+1$ to n.

Problem 2.29   If $X \sim$ BINO(n, p), show that the lines $y = x \Pr(x)/\Pr(x-1)$, and the line $y = mx + c$ intersect if $c = (n+1)p/q$ and $m = -p/q$, where $q = 1 - p$.

Problem 2.30   If $X_1, X_2, \ldots, X_N$ are IID exponential random variables where $N \sim$ BINO(n, p), find the distribution of $S = X_1 + X_2 + \cdots + X_N$.

2.8   SPECIAL BINOMIAL DISTRIBUTIONS

There are several extensions of the binomial distribution available in the literature. The popular among them are truncated binomial distributions, inflated binomial distributions, success-run distributions, distribution of the difference between successes and failures, etc. [104] Other popular extensions include the quasi-binomial distributions ([19], [21]) with PMF $f(x) = \theta/(1+\theta)^{n}\binom{n}{x}p^{x-1}q^{n-x}$ where $p = \theta + \lambda x$, $q = 1 - \lambda x$, $0 \le \lambda < 1/n$, $0 < \theta < \infty$, and the generalized binomial distribution of Jain and Consul [35], and Consul and Shenton [22] with PMF $f(x) = (s/t)\binom{t}{x}p^{x}q^{t-x}$ where $t = s + rx$, $x = 0, 1, 2 \ldots, 0 < p < 1/r$ and r, s are positive integers.

2.8.1   DISTRIBUTION OF $U = |X - Y|/2$

The number of successes and failures in a binomial distribution is related through n. If there are x successes, there exist $n - x$ failures and *vice versa*. In other words, they must add up to the total number of trials. The distribution of $U = |X - Y|/2$ finds applications in various fields like epidemiology, stochastic processes and games. Let X denote the number of successes (or Heads), and Y denote the number of failures (or Tails) in n independent Bernoulli trials with the same probability of success p. Obviously U takes the values $(0, 1, \ldots, n/2)$. If n is even, it can take the value 0 in just one way—when both X and Y are $n/2$. The probability of this case is $\binom{n}{\frac{n}{2}}p^{n/2}q^{n-n/2} = \binom{n}{\frac{n}{2}}(pq)^{n/2}$. There exist two ways in which all other values are materialized. First consider the number of successes exceeding the number of failures by x. Let t be the number of failures (so that the number of successes is $t + x$). Then $t = (n - x)/2$, and $t + x = (n + x)/2$. The probability of this happening is

$$f_u(x; n, p) = \binom{n}{\frac{n+x}{2}} p^{\frac{n+x}{2}} q^{\frac{n-x}{2}} = (pq)^{n/2}\binom{n}{\frac{n+x}{2}}(p/q)^{x/2}, \qquad (2.68)$$

for $x = 2, 4, \ldots, n$. Next consider the number of failures exceeding the number of successes by x.

Let t be the number of failures. Then $t = (n + x)/2$, and $t - x = (n - x)/2$. The probability of this happening is

$$f_u(x; n, p) = \binom{n}{\frac{n-x}{2}} p^{\frac{n-x}{2}} q^{\frac{n+x}{2}}. \tag{2.69}$$

Using $\binom{n}{x} = \binom{n}{n-x}$ this becomes

$$f_u(x; n, p) = \binom{n}{\frac{n+x}{2}} p^{\frac{n-x}{2}} q^{\frac{n+x}{2}} = (pq)^{n/2} \binom{n}{\frac{n+x}{2}} (q/p)^{x/2}, \tag{2.70}$$

for $x = 2, 4, \ldots, n$. Adding (2.69) and (2.70) gives the probability of U assuming the value u as

$$f_U(u; n, p) = \begin{cases} \binom{n}{n/2+u}(pq)^{\frac{n}{2}} [(p/q)^u + (q/p)^u] & \text{for } u = (1, \ldots, n/2); \\ \binom{n}{\frac{n}{2}}(pq)^{n/2} & \text{for } u = 0; \\ 0 & \text{otherwise.} \end{cases}$$

Putting $t = n/2 + u$, this can also be written as

$$f_T(t; n, p) = \begin{cases} \binom{n}{t} [q^n(p/q)^t + p^n(q/p)^t] & \text{for } t = (n/2 + 1, \ldots, n); \\ \binom{n}{\frac{n}{2}}(pq)^{n/2} & \text{for } t = n/2; \\ 0 & \text{otherwise.} \end{cases}$$

Take q^n as a common factor, and write $(q/p)^t = (p/q)^{-t}$ to get the alternate form

$$f_T(t; n, p) = q^n \binom{n}{t} [(p/q)^t + (p/q)^{n-t}]. \tag{2.71}$$

For $p = q$, this simply becomes $\binom{n}{t}/2^{n-1}$. A similar result could be derived when n is odd. See Table 2.4 for some sample values.

There is another way to derive the above distribution. Let there be x successes and $(n - x)$ failures in n trials. Then $S - F = x - (n - x) = 2x - n = y$ (say) where S denotes the successes and F denotes the failures. Clearly, y takes the values $-n, -n + 2, \ldots, 0, \ldots, n - 2, n$:

$$P(Y = y) = P(2x - n = y) = P(x = (n + y)/2) = \binom{n}{(n + y)/2} p^{(n+y)/2} q^{(n-y)/2}. \tag{2.72}$$

The distribution of $|Y|$ is $f(y) + f(-y)$. Put $y = -y$ in (2.72) and add to get

$$f(y) = \binom{n}{(n + y)/2} p^{(n+y)/2} q^{(n-y)/2} + \binom{n}{(n - y)/2} p^{(n-y)/2} q^{(n+y)/2}. \tag{2.73}$$

Table 2.5: Distribution of $U = |X - Y|/2$ for n even

$x \backslash (n, p)$	(6, 0.2)	(10, 0.6)	(20, 0.3)	(20, 0.9)	(20, 0.5)
0	0.0819	0.2007	0.0308	0.0000	0.17620
1	0.2611	0.3623	0.0774	0.0001	0.32040
2	0.3948	0.2575	0.1183	0.0004	0.24030
3	0.2622	0.1315	0.1653	0.0020	0.14790
4		0.0419	0.1919	0.0089	0.07390
5		0.0062	0.1789	0.0319	0.02960
6			0.1304	0.0898	0.00920
7			0.0716	0.1901	0.00220
8			0.0278	0.2852	0.00040
9			0.0068	0.2702	.000038
10			0.0008	0.1216	.000002
Sum	1.0000	1.0000	1.0000	1.0000	1.0000

Second column onward give values of n and p. As u varies between 0 and $n/2$, there are $n/2 + 1$ values in each column.

Use $\binom{n}{x} = \binom{n}{n-x}$ and take common factors outside to get the above form. This result has important application in several fields. Consider a problem of modeling the spread of an epidemic like novel coronavirus (COVID-19). Suppose the total population is divided into two sets S1 (of infected people) and S2 (of uninfected people). Uninfected people may move to the S1 group (when infected), and infected people may move to S2 (when they recover completely). An epidemiologist may want to model this transition on a fixed time interval (say day-to-day basis) using a sample of size n (X infected, $n - X$ non-infected people). Likewise, we can model passive smoke hazard rate among X smokers and $n - X$ non-smokers in an enclosure (office, theater, etc.) so that the difference between them can give us information on secondary smoke hazard. The above result can tell us the change in probability so that remedial measures can be taken to decrease the spread of an epidemic. Similarly, some species produce offspring in a fixed ratio (this is roughly 50–50 for humans (the gender ratio M/F differs among various regions), but more females than males are born for several animals like cows, elephants, etc.). Our variable of interest is the difference between male and female progeny born in a particular time interval (say one year). If this probability is known from historical data, we could predict variations from that expected in a particular future time using the above formula.

2.8.2   TRUNCATED BINOMIAL DISTRIBUTIONS

Truncated distributions find applications in various fields. There are several processes where either the zero-count is not of use, or the observational apparatus starts counting only for non-zero event occurrences. The zero-truncated binomial (ZTB) distribution discards the $x = 0$ value so that the total probability is reduced by q^n. It finds applications where an excessively large count is observed for $x = 0$. Classical (un-truncated) models will under-predict such values. Hence, it is called "excess-zeros" problem. There exist three popular approaches to "excess-zeros" problem:—(i) zero-truncated models, (ii) zero-inflated models, and (iii) hurdle models. Even if the zero-count (namely $x = 0$) is discarded, the rest of the values may be over-dispersed so that the estimates of the parameters are biased. The zero-inflated models discussed below are suitable in such situations.

The PMF of ZTB distribution is

$$f(x; n, p) = \binom{n}{x} p^x q^{n-x} / (1 - q^n), \quad 0 \leq q = 1 - p \leq 1 \quad \text{for} \quad x = 1, 2, \ldots, n. \quad (2.74)$$

As $q^n = F(0)(1 - F(0) = S(0))$ an alternate representation is $f(x; n, p) = \binom{n}{x} p^x q^{n-x} / (1 - F(0)) = \binom{n}{x} p^x q^{n-x} / S(0)$, where $F(x)$ denotes the CDF, and $S(x)$ the survival function.

The PGF can be found easily as

$$P_x(t) = \sum_{x=1}^{n} t^x \binom{n}{x} p^x q^{n-x} / (1 - q^n). \quad (2.75)$$

Take $(1 - q^n)$ outside the summation and combine t^x and p^x to get

$$P_x(t) = 1/(1 - q^n) \sum_{x=1}^{n} \binom{n}{x} (pt)^x q^{n-x}. \quad (2.76)$$

The sum can be written as $(q + pt)^n - q^n$. This gives the PGF of ZTB distribution as $P_x(t) = [(q + pt)^n - q^n]/(1 - q^n)$. Similarly, the MGF of ZTB follows as $M_x(t) = [(q + p \exp(t))^n - q^n]/(1 - q^n)$.

Left-truncated binomial distribution is obtained by truncating at a positive integer K. The resulting PMF is

$$P_x(t) = \binom{n}{x} p^x q^{n-x} / \left(1 - \sum_{x=0}^{k} \binom{n}{x} p^x q^{n-x} \right) \quad \text{for} \quad x = k + 1, \ldots, n. \quad (2.77)$$

2.8.3   INFLATED BINOMIAL DISTRIBUTIONS

The ZTB distribution gives a better fit for "excess-zeros" problem. We could break the count for $x = 0$ as "true zeros" and "excess zeros." In a stock-market trading scenario, the zero count for a share may be those traders who never bought that particular share plus those traders who have bought that share at least once in the past but did not buy it on the particular trading day. The zero-inflated binomial (ZIB) distribution is a better alternative for over-dispersed data. Unlike the ZTB distribution, the ZIB model considers zero-counts using a weighting parameter:

$$f(x; n, p, \theta) = \begin{cases} 1 - \theta & \text{if } x = 0, \\ \theta g(x; n, p, \theta) & \text{if } x > 0, \end{cases}$$

where $g(x; n, p, \theta)$ is the ZTB distribution.

2.8.4 LOG-BINOMIAL DISTRIBUTION

This distribution finds applications in high-voltage and thermal engineering. If $X = \exp(Y)$ has PMF $f(x) = \binom{n}{x} p^x q^{n-x} / [(1 - q^n)]$, the distribution of Y is $f(y) = \binom{n}{e^y} p^{e^y} q^{n-e^y} / [(1 - q^n)]$ for $y = 0 = \ln(1), \ln(2), \ldots, \ln(n)$. If $X = \log(Y)$ is $ZTB(n, p)$ we say that X has a log-binomial distribution.[1] As $f(x) = \binom{n}{x} p^x q^{n-x}$, the distribution of Y is given by $f(y) = \binom{n}{\ln(y)} p^{\ln(y)} q^{n-\ln(y)} / [(1 - q^n)]$ for $y = 1, \exp(1), \exp(2), \ldots, \exp(n)$.

2.9 APPLICATIONS

The binomial distribution is used to model situations that give rise to counts of dichotomous data and in inference relating to such data. It finds applications in almost all branches of engineering and the applied sciences. The following pages list some of the more popular applications.

2.9.1 MICROBIOLOGY

Microbiology uses a technique called "assaying" to find the concentration of microorganisms in laboratory samples. Different techniques are used depending on whether the sample is soluble (say in water or another liquid medium) or not. One simple technique uses a geometric dilution factor to get dilutes in the proportions $1/2, 1/2^2, 1/2^3, 1/2^4$, etc., as shown in Table 2.6. This is called limiting dilution assay and is discussed in Chapter 6. For water-soluble sample, this means that we prepare 1:2, 1:4, 1:8, etc., dilutions. Each of the n obtained dilutes are labeled. Multiple samples are then taken from each of the dilutes. A detection test is applied to see if microbes are present or not. Assume that a stage is reached beyond which the microbes are absent in some samples of the assays. Then the assay at the cutoff between presence and absence of microbes is marked off. Suppose we scan the table row-wise from top to bottom. Presence of the microbe is indicated by a 1 and absence by 0. Note down the first row k in which there is at least one column (sample) with a 0 (microbe absent). In other words, at least one of the jth replicate at ith dose is zero, binomial distribution is fitted to that row onward until all zero rows

---

[1]Note that $y = 0$ is discarded as $\log(0) = -\infty$.

Table 2.6: Diluted assay and binomial distribution

Dilution	0|1	0|1	0|1	0|1	0|1	0|1	0|1	0|1	0|1	0|1	Comments
θ	1	1	1	1	1	1	1	1	1	1	
$\theta/2$	1	1	1	1	1	1	1	1	1	1	
$\theta/2^2$	1	1	1	1	1	1	1	1	1	1	
$\theta/2^3$	1	0	1	1	0	1	1	1	1	1	Binomial
$\theta/2^4$	0	1	1	0	1	0	1	1	1	0	Binomial
$\theta/2^8$	1	0	1	1	0	0	1	1	0	0	Binomial
$\theta/2^{16}$	0	0	1	1	0	0	1	0	0	0	Binomial
$\theta/2^{32}$	0	0	0	0	0	$\cdot$ 0	0	0	0	0	

are encountered. Binomial distribution is applicable from that row onward, until a row with all zeros are encountered. In other words, every row that contains 1's and 0's can be modeled using binomial distribution because the probability that microbe is present remains the same (since concentration is the same). Thus, rows 3, 4, 5, corresponding to dilutions $\theta/2^3$, $\theta/2^4$, and $\theta/2^5$ can be modeled using binomial distribution. As dilution decreases steadily, the Poisson approximation to binomial distribution is applicable for rows with few 1's (microbes present in few samples) when n is very large. If m samples were taken from that particular dilute, chances are that some of these samples will contain the microbes and others will not, so that a binomial distribution is applicable. Let X denote the number of samples containing microbes at the cutoff assay. As each of the m samples were drawn from an assay with the same concentration, it is safe to assume that p, the probability of finding the microbe is a constant (but p is not a constant for assays with varying concentrations).

2.9.2   EPIDEMIOLOGY

Suppose a large population under study is divided into a 2×2 table according to a factor of interest like positive and negative for a particular disease or condition along columns and a factor along rows. The rows may correspond to a current classification (say in a retrospective study) of past data. For convenience, the entries can be scaled to proportions as shown in Table 2.6. A question of natural interest is to find out the association between factors and diseases.

2.9.3   GENETICS

Many properties encountered in genetics are dichotomous. Examples are occurrences of mutations in a gene resulting in offspring of different traits, exterior color, shape, and size of produce (like peas, beans, etc.). If the probability of occurrence of a trait is known from past data, the frequency of occurrence of a trait can be modeled using binomial distribution. In some applications like resistance mutations where one of the probabilities is very small, a Poisson distribution (Chapter 6) may be used to compute the desired probabilities.

2.9.4   RELIABILITY

Consider a population of N independent and identical items with CDF $F(t)$ (failure distribution function in reliability engineering). Any of the units may fail to function during its operation. The order in which units fail is unimportant in the modeling process, but it is just how many of them failed is important to us. If $N_x(t)$ denotes the number of items in operation at time t, we could express it in terms of binomial distribution as follows:

$$\Pr[N_x(t) = y] = \binom{N}{y} p^y (1-p)^{N-y}, \tag{2.78}$$

where $p = E[N_x(t)]/N$.

Problem 2.31   Ten identical aircraft engines are put on a continuous stress test to see if they can function continuously for 40 h. If the probability that any of them will fail is 1 in 1,000, what is the probability of at most 2 failures?

Age-dependent failure rate processes are those in which the reliability decreases over time.[2] One example is the aircraft engines. Often, a thorough overhaul (maximal repair) of the equipment will restore it to a new condition. Most repairable equipments and machinery exhibit the *increasing failure rate*, meaning that they may fail time and again with shorter durations even after maximal repair (unless they are totally replaced with new ones). Examples are induction cookers, refrigerators, engines, and landing gear of aircrafts. The Poisson SF (with parameter n/λ_p where λ_p is the mean inter arrival time) has been used to model such machinery by providing an upper bound on the probability of n or more failures in $[0, t)$.

2.9.5 K-OUT-OF-n SYSTEM OPERATION

K-out-of-n systems find applications in several fields like electronics and hydrolic engineering, ensemble machine learning methods, etc. When $k = 1$ we get the 1-out-of-n system, which is a parallel system. When $k = n$ we get n-out-of-n system, which is simply a series system. It is assumed in the following discussion that k is neither 1 nor n, and that all components are identical and independent. Consider a system comprising of n units. This system will work successfully if any of the $k < n$ systems work. If the probability (reliability) of each component

---

[2]The opposite is the age-persistent failure rate processes in which a minimal repair will restore it to working conditions. Examples are most of the electrical equipments and some electronic equipments with simple circuitry like radios, sound amplifiers, etc.

is p, the probability of system operating successfully is

$$\sum_{j=k}^{n} \binom{n}{j} p^j q^{n-j}. \tag{2.79}$$

If the time to failure is exponentially distributed, we have $p = \exp(-j\lambda t)$. A similar concept is the N-modular redundant (NMR) systems in which N identical units that work independently are connected in parallel, and a *voter* selects the correct output from among them. The NMR voter is programmed to select the correct output from more than half of the units. Assume for simplicity that $N = 2m + 1$ is odd. If the voter does not fail, the system is operational when m or more units work, in which case the reliability is found as

$$R = \sum_{j=m}^{N} \binom{N}{j} p^j q^{n-j}, \tag{2.80}$$

where p is the reliability of each unit.

2.9.6 K-OUT-OF-n DETECTION

Modern radar systems use multiple detection tries to conclusively identify a target. This lessens probability of false alarms (PFA) and increases probability of detection (PD). Suppose n independent tries are made and m out of them detects a target. Then if p is the probability of a threshold crossing in a single trial, the required probability is given by SF of BINO(n, p) as $\sum_{x=m}^{n} \binom{n}{x} p^x q^{n-x}$.

2.9.7   ENSEMBLE METHODS

The literal meaning of an "ensemble" is *a collection of items or things that works as a whole*. It is a cohesive model to which many loosely coupled base models contribute their individual knowledge. Ensemble models is an emerging research topic with applications in machine learning, data analysis (regression, clustering), multi-dimensional scaling, weather forecasting, price, and index forecasting in econometrics, management, etc. It is called "mixture of experts model" in some engineering fields. A part of the variance in any model built from static data is attributed to the choice of the training set. By combining models built from different training data results in reduced error variance. This is the guiding principle of ensemble models.

Definition 2.32   An ensemble model is a single logical model built by combining two or more loosely coupled base models that aims to improve the robustness and accuracy by leveraging the power of its components in an optimal way.

When used for classification, it combines two or more competing models into a *committee ensemble model*. Ensemble learning models (ELM) used in classification algorithms are formed by combining two or more classifiers in an optimal way. The simplest ensemble model uses n identical and independent sub-models which are called weak classifiers. A voting method is usually employed to compare the performance of the model. If X_k denote a random variable that acts as an indicator function (meaning that $X_k = 1$ if kth ensemble correctly classifies the data), it is BER(p) where p is the probability of prediction accuracy of each sub-model. If majority voting is used, the variable $Y = \sum_{k=1}^{n} X_k$ is BINO(n, p). Let $n = 2m + 1$, an odd integer. Then the prediction accuracy of the ensemble can be expressed as the SF of BINO(n, p) as

$$P_n = \sum_{k=m+1}^{n} \binom{n}{k} p^k q^{n-k}. \tag{2.81}$$

2.9.8   SQC

Most of the modern manufacturing industries have strict quality control of produced items. Sampling schemes are used to examine the quality of products (called lots). Producers are concerned with the probability that a lot produced when a process is under control will get rejected (called producers risk). Consider a fixed time interval T during which N independent units are produced. If n items are selected at random in the inspection sample from N produced items, and process fraction defective is p, we could find producers risk if it is assumed that p remains a constant. If a lot of size n is rejected if there are c or more defectives in it, the producers risk is given by the SF of BINO(n, p) as $\sum_{x=c+1}^{n} \binom{n}{x} p^x (1 - p)^{n-x}$.

Control Charts for Nonconforming Fractions
Control chart is a statistical technique that depict patterns of quality variations in manufactured items. As control charts are based upon counts of items that do not conform to a set standard, the binomial distribution is widely used to construct such charts. If p denotes the probability that a randomly manufactured (or assembled) unit does not conform to a set standard, the number of units x that do not conform to the standard out of n unit produced has a BINO(n, p) distribution. From this, the sample fraction can be found as $\hat{p} = x/n$, the mean and variance of which are, respectively, p and pq/n. Now the LCL and UCL are found.

The α risk for nonconforming items is then the SF of binomial random variable BINO(n, p_0) where p_0 is the fraction defective when the process is in control, c is the acceptance number, and β risk is the CDF up to c of BINO(n, p_1) where p_1 is the fraction defective when the process is out-of-control.

OC Curves
Operating characteristic (OC) curve is a visual tool that presents the risks involved in acceptance sampling. Probabilities of accepting lots that contain varying levels of defectives are plotted as a continuous curve with the X-axis representing the percent of non-conforming items and Y-

axis representing the probability of acceptance. Depending on whether a process is discrete or continuous, two types of OC curves (called Type-A and Type-B) are used. Type-A process usually uses the hypergeometric distribution whereas Type-B process uses binomial (or Poisson as an approximation to binomial[3]) distribution. For every fraction of non-conforming items, we generate the CDF of binomial (or Poisson) from 0 to r where r is the total number of nonconformances in a sample. Usually, the fraction defectives is restricted to an upper bound (say 10% or $p = 0.10$).

2.9.9   ELECTRONICS

Binomial distribution has been applied in semiconductor industry to model defects on wafers when defects occur independently and probability of occurrence remains the same (which is strictly speaking unrealistic, but could give good approximations). A wafer holds multiple identical "dies." Consider a wafer with N dies and suppose it has n defects in total. The average defects per die is $\lambda = n/N$, which is modeled using the Poisson law (Chapter 6). Defects on each die is modeled using the binomial distribution $BINO(n, p)$. Probability that k defects land on a die is given by the PMF of $BINO(k; n, p)$.

Example 2.33   Malfunctioning electronic device   Consider an electronic device containing n transistors from the same manufacturer. The probability of each transistor malfunctioning is known from previous observations over a long period of time to be p. Find the probability that (i) at most three transistors malfunction, and (ii) none of the transistors malfunction.

Solution: We assume that the transistors malfunction independent of each other. Then the number of transistors that malfunction has a binomial distribution. Hence, the required probability is $\Pr[X \leq 3] = \sum_{x=0}^{3} \binom{n}{x} p^x q^{n-x}$. Probability that none of them malfunction is $\binom{n}{0} p^0 q^{n-0} = q^n$.

The binomial distribution is also used in *batch testing* of multiple components simultaneously. Consider a batch of n identical items or components to be tested simultaneously. If the lifetime of each item is exponentially distributed, the probability of failure in time T is $p = 1 - \exp(-\lambda T)$, so that the probability of k failures is given by $BINO(n, 1 - \exp(-\lambda T))$ as

$$p = \binom{n}{k} [1 - \exp(-\lambda T)]^k [\exp(-\lambda T)]^{n-k}. \tag{2.82}$$

Problem 2.34   A food packaging company uses automated machines to fill cartons of food packets. From past data, it is found that the probability that a food packet will be under-filled

---

[3]See discussion of this approximation in Chapter 6.

is 1/80. In a shipment containing 20 food packets, what is the probability of finding (i) exactly 2 under-filled packets, and (ii) between 2 and 5 such packets?

Problem 2.35   An aircraft has six identical tires. From past experience, it is known that tire bursts occur once in 820 landings. The aircraft can still be safely landed if the number of tire bursts upon landing are two or less. What is the probability of an unsafe landing?

Problem 2.36   Find the mean and variance of the reparametrized version of binomial distribution obtained by putting $p/q = \theta$ as

$$\binom{n}{x}[\theta/(1+\theta)]^x[1/(1+\theta)]^{n-x}, \quad \theta > 0. \tag{2.83}$$

Problem 2.37   If $X \sim BINO(n, p)$ and $Y|X \sim BINO(X, q)$ then prove that the unconditional distribution of Y is $BINO(n, pq)$.

Problem 2.38   Water used in a nuclear power plant is let out into 25 quadrats. Different types of marine life are farmed in the quadrats to see if it is harmful to them. From routine inspections it is found that 6 out of 25 quadrats are harmful to some species. If 8 quadrats are examined, what is the probability that 4 or more of them are harmful to marine life?

2.10   SUMMARY

This chapter introduced the Bernoulli and binomial distributions, and their basic properties. It also discussed tail areas of binomial distribution using incomplete beta function. Recurrence relationships satisfied by the PMF and moments of binomial distribution are also discussed. Special binomial distributions like truncated, inflated, and log-binomial distributions are briefly introduced. The chapter ends with a list of applications of binomial distribution in engineering and scientific fields.

CHAPTER 3

Discrete Uniform Distribution

> **After finishing the chapter, readers will be able to ...**
>
> - Understand discrete uniform distributions.
> - Describe properties of discrete uniform distribution.
> - Apply the discrete uniform distribution in practical problems.

3.1   INTRODUCTION

Each event in a sample space has an equal chance of occurring when the outcomes of a random experiment are equally likely. One example is the throw of an unbiased coin or die (with six faces). Although this kind of events have theoretical importance, they are rarely used in practice because most of the practical phenomena exhibit some kind of randomness. However, they find applications in sampling with replacement, occupancy type problems, etc.

3.2   DISCRETE UNIFORM DISTRIBUTION

A random variable that takes equal probability for each of the outcomes has a discrete uniform distribution (DUNI[N]). The PMF is given by

$$f(x) = \Pr[X = k] = 1/N, \quad \text{for} \quad k = 1, 2, \ldots, N \quad \text{where} \quad N > 1. \tag{3.1}$$

Thus, we get DUNI(N) when there exist finitely many possible equally likely outcomes. As the probabilities should sum to one, this distribution always has finite range. Alternately, $f(x) = \Pr[X = k] = 1/|\Omega|$ for $x \in \Omega$, if the entire sample space is denoted by Ω, and $|\Omega|$ denotes the cardinality. The variate values are equi-spaced positive integers in most applications (either starting with 0 or with 1). They are often consecutive integers, although this is not a theoretical requirement. For $N = 2$, this becomes the Bernoulli distribution with $p = 1/2$. As each of the probabilities are equal, $f(x)/f(x + k) = 1$ for all k in the range. It is also called discrete rectangular distribution.

A displaced discrete uniform distribution DUNI$[a, b]$ (where $b > a$) can be defined as $f(x) = \Pr[X = a + k] = 1/N$, for $k = 0, 1, 2, \ldots, b - a$ (or $f(x) = \Pr[X = k] = 1/N$ for $k = a, a + 1, a + 2, \ldots, b - 1$). Choosing $a = 0, b = N$ gives another form of the distribution as $f(x) = 1/(N+1)$, for $x = 0, 1, 2, \ldots, N$. We could shift the origin by c (positive or negative) to get the more general DUNI(N) as $f(x) = \Pr[X = k] = 1/N$, for $k = c, c+1, \ldots, c + N - 1$. A change of origin and scale transformation can be used to get the general rectangular distribution (which may assume non-integer values). An alternate representation is

$$f(x) = \Pr[X = k] = 1/(2\mu - 1), \quad \text{for} \quad k = 1, 2, \ldots, N, \tag{3.2}$$

where $N > 1$ and $\mu = (N+1)/2$ (see below).

3.2.1   MEAN AND VARIANCE

The mean is $E(X) = \sum_{x=1}^{N} x/N = (1/N)(1 + 2 + \cdots + N) = N(N+1)/(2N) = (N+1)/2$. The mean is an integer when N is odd, and an "half-integer" for N even. Thus, the DUNI(N) random variable can never take the mean value for N an even number. To find the variance, we first find $E(X^2)$.

$$E(X^2) = (1/N)\left[1^2 + 2^2 + \cdots + N^2\right] = N(N+1)(2N+1)/(6N) = (N+1)(2N+1)/6.$$

From this we get variance as $V(X) = (N+1)(2N+1)/6 - (N+1)^2/4$. Take one $(N+1)$ as common factor, and 12 as LCM to get $(N+1)[2(2N+1) - 3(N+1)]/12 = (N+1)(N-1)/12$. Thus, the mean and variance are $\mu = (N+1)/2$ and $V(X) = \sigma^2 = (N^2 - 1)/12$. The variance is $1/4$ for $N = 2$, $2/3$ for $N = 3$, and greater than one for all N values ≥ 4. If the domain is taken as $\{0, 1, 2, \ldots, N\}$, the PMF becomes $f(x; N) = 1/(N+1)$. This has mean $N/2$ and variance $N(2N+1)/6 - N^2/4 = N^2/12 + N/6 = N(N+2)/12 = \mu(N+2)/6$.

Example 3.1   CDF of DUNI(N) distribution   Find the CDF of DUNI[N], and obtain the mean using $E(X) = \sum_k P(X \geq k)$.

Solution: Assume the PMF is $f(x) = \Pr[X = k] = 1/N$. The CDF $F(x) = P[X \leq x] = \sum_{k=1}^{x} 1/N = x/N$. This can be expressed in terms of indicator functions as $F(x) = (1/n)\sum_{j=1}^{n} \mathbb{I}_{x_j \leq x}$. From this we get $P(X > x) = 1 - [x/N]$ and $P(X \geq x) = 1 - [(x - 1)/N]$. Now $E(X) = \sum_k P(X \geq k) = \sum_{k=1}^{N}(1 - [(k-1)/N]) = \sum_{k=1}^{N} 1 - 1/N \sum_{k=1}^{N}(k - 1) = N - (1/N)[1 + 2 + 3 + \cdots + (N-1)] = N - (1/N)(N-1)N/2$. This simplifies to $E(X) = \mu = (N+1)/2$.

Example 3.2   MD of DUNI(N) distribution   Find the MD of DUNI[N] distribution using the Power method (Chapter 1).

Solution: We know that the mean of DUNI[N] distribution is $\mu = (N+1)/2$. Assume that $N = 2m + 1$ is odd, so that the summation is carried out to $\mu - 1 = m$. Using theorem in

Chapter 1, the MD is given by

$$\text{MD} = 2 \sum_{x=ll}^{\mu-1} F(x) = 2 \sum_{x=1}^{m} x/N. \tag{3.3}$$

Take $(1/N)$ outside the summation and evaluate $\sum_{x=1}^{m} x = m(cm+1)/2$. This gives

$$\text{MD} = (2/N) * m(m+1)/2 = m(m+1)/N. \tag{3.4}$$

Substitute $m = (N-1)/2$ to get $\text{MD} = ((N-1)/2)((N+1)/2)/N = (N^2-1)/(4N)$. Next assume that $N = 2m$ is even, so that the summation is carried out to $\mu - 1 = m$, which is the nearest integer. This gives

$$\text{MD} = 2 \sum_{x=ll}^{m} F(x) = 2 \sum_{x=1}^{m} x/N. \tag{3.5}$$

This simplifies to $\text{MD} = m(m+1)/N$. Put $m = N/2$ to get $\text{MD} = (N+2)/4$.

Problem 3.3 If $X \sim \text{DUNI}(N)$, find the distribution of $Y = X - (N+1)/2$, and its mean.

Problem 3.4 Find the value of N in $\text{DUNI}(N)$ for which $\mu : \sigma^2 = 3 : 2$.

Problem 3.5 Find the least value of N in $\text{DUNI}(N)$ for which the variance is greater than the mean.

3.3 PROPERTIES OF DISCRETE UNIFORM DISTRIBUTION

This distribution has a single parameter. If $X \sim \text{DUNI}(N)$, with support $(0, 1, \ldots, N)$ then $Y = N - X$ is identically distributed (if the support is $(1, \ldots, N)$, then $Y = N + 1 - X$). The hazard function is $f(x; N)/S(x; N) = (1/N)/(1 - x/N) = 1/(N - x)$. The MGF is easy to find as

$$E\left(e^{tx}\right) = \sum_{x=1}^{N} e^{tx}/N = \frac{1}{N}\left[e^t + e^{2t} + \cdots + e^{Nt}\right] = \frac{e^t}{N}\frac{1 - e^{Nt}}{1 - e^t} \tag{3.6}$$

for $t \neq 0$ and $= 1$ for $t = 0$. The PGF is obtained by replacing e^t in the above by t as $P_x(t) = \frac{t}{N}\frac{1 - t^N}{1 - t}$. The characteristic function is written as $\phi_x(t) = (1 - e^{itN})/[N(e^{-it} - 1)]$, where we have divided both numerator and denominator by e^t.

Table 3.1: Properties of discrete uniform distribution

Property	Expression	Comments
Range of X	$x = 1,...,N$	Discrete, finite
Mean	$\mu = (N + 1)/2$	Int for N odd
Variance	$\sigma^2 = (N^2 - 1)/12 = \mu * (\mu - 1)/3$	$\Rightarrow \mu > \sigma^2$ for $N < 7$
Mode	any x	
Skewness	$\gamma_1 = 0$	Special symmetry
Kurtosis	$\beta_2 = \frac{3}{5}(3 - \frac{4}{N^2-1})$	
CV	$\{(N - 1)/[3(N + 1)]\}^{1/2}$	
MD	$(1 + \lfloor(N+1)/2\rfloor)\lfloor(N+1)/2\rfloor/N$	$(N^2 - 1)/4N$
Moments	$\mu'_r = (1^r + 2^r + ... + N^r)/N$	Bernoulli number
MFG	$e^t(1 - e^{Nt})/[N(1 - e^t)]$	
$\phi_x(t)$	$(1 - e^{itN})/[N(e^{-it} - 1)]$	
PFG	$[t(1 - t^N)]/[N(1 - t)]$	
Recurrence	$f(x;N)/f(x - 1;N) = 1$	$f(x) = f(x - 1)$
SF	$\sum_{x=k}^{N} 1/N = (N - k + 1)/N$	

Truncation results in the same distribution with higher probability for each x value.

The square of the coefficient of variation is then $[(N + 1)(N - 1)/12]/[(N + 1)/2]^2$. Cancel out common factors to get $CV^2 = (N - 1)/[3(N + 1)]$. Now take positive square root to get the CV as $(1/\sqrt{3})[(N - 1)/(N + 1)]^{1/2}$. Other properties are given in Table 3.1.

Example 3.6 Equal mean and variance for the DUNI(N) Find the value of N for which a DUNI(N) distribution has (i) the same mean and variance, and (ii) variance is twice the mean.

Solution: We know that the mean is $(N + 1)/2$ and variance is $(N^2 - 1)/12$. Equate and cross-multiply to get $6(N + 1) = N^2 - 1$ or equivalently $N^2 - 6N - 7 = 0$. This being a quadratic in N has two roots. Choose the positive root to get $N = 7$. Thus, the mean ($= 4$) coincides with the variance for $N = 7$. Proceed as above for part (ii) to get $2 * (N + 1)/2 = (N^2 - 1)/12$. This results in $N^2 - 12N - 13 = 0$. The roots are $(12 \pm 14)/2$. The positive root is $26/2 = 13$. Thus, the variance ($= 14$) is twice the mean ($= 7$) for $N = 13$.

Problem 3.7 Find the value of N for which the variance of a DUNI(N) distribution is thrice the mean.

3.3.1   SPECIAL SYMMETRY

This distribution has a special symmetry (called *flat symmetry*) as all probabilities are equal. Hence, the median coincides with the mean, and the skewness γ_1 is zero. The coefficient of kurtosis is $\beta_2 = \frac{3}{5}(3 - \frac{4}{N^2-1})$. This shows that it is always platykurtic.

Truncated discrete uniform distributions are of the same type with the probabilities simply enlarged (because dividing by the truncated sum of probabilities simply enlarges each individual probability).

Example 3.8   Variance as a function of μ for the DUNI(N)   Express the variance of DUNI(N) as a function of μ alone.

Solution: We know that $\mu = (N + 1)/2$ and $\sigma^2 = (N^2 - 1)/12$. From $\mu = (N + 1)/2$ we get $N = 2 * \mu - 1$. Write $\sigma^2 = (N^2 - 1)/12 = (N - 1)(N + 1)/12$. Substitute $(N + 1)/2 = \mu$ to get $\sigma^2 = [(2 * \mu - 1) - 1] * \mu/6 = (\mu - 1) * \mu/3$. If the variance of a DUNI(N) distribution is estimated from data, we can obtain an estimate of the mean as follows. Write the above as a quadratic equation $x^2 - x - 3k = 0$ where $x = \mu$ and $k = \sigma^2$. This has positive root $x = (1 + \sqrt{1 + 12 * k})/2$. Put the values for x and k to get $\mu = (1 + \sqrt{1 + 12 * \sigma^2})/2$. Alternatively, we could first estimate N from variance as $N = \sqrt{1 + 12 * \sigma^2}$ and obtain the mean as $(N + 1)/2$.

Example 3.9   Factorial moments of DUNI(N)   Find the factorial moments of DUNI[N], and obtain the mean.

Solution: By definition

$$\mu_{(k)} = E[X(X - 1)\ldots(X - k + 1)] = \sum_{x=1}^{N} x(x - 1)\ldots(x - k + 1)(1/N). \qquad (3.7)$$

As $(1/N)$ is a constant while summing w.r.t. X, take it outside the summation and adjust the indexvar to vary from k to N. This gives $\mu_{(k)} = (1/N) * \sum_{x=k}^{N} x(x - 1)\ldots(x - k + 1)$. Multiply and divide by $1.2\ldots.(x - k)$ and write this as $\mu_{(k)} = (1/N) * \sum_{x=k}^{N} x!/(x - k)!$. Next multiply and divide by $k!$ and write $x!/[k!(x - k)!]$ as $\binom{x}{k}$. The RHS becomes $(k!/N) * \sum_{x=k}^{N} \binom{x}{k}$. Now use $\sum_{x=k}^{N} \binom{x}{k} = \binom{N+1}{k+1} = (N + 1)!/[(k + 1)! * (N - k)!]$. Write $(k + 1)!$ in the denominator as $(k + 1) * k!$ and cancel out the $k!$ to get

$$\mu_{(k)} = (1/N(k + 1)) * (N + 1)!/(N - k)!. \qquad (3.8)$$

Put $k = 1$ to get the mean as $\mu = 1/(2 * N)[(N + 1)!/(N - 1)!] = (N + 1)/2$.

Example 3.10   Distribution of $U = X + Y$.   If X and Y are IID DUNI(N) with the same range, find the distribution of $U = X + Y$.

Solution: Without loss of generality, we assume that X and Y takes values $1, 2, \ldots, N$. Then $\Pr[X + Y = k] = \Pr[X = t \cap Y = k - t] = \Pr[X = t] \cap \Pr[Y = k - t]$ due to independence. As t is arbitrary, this becomes $\sum_{t=1}^{k-1} 1/N^2 = (k-1)/N^2$. Hence, $f_u(u) = (u-1)/N^2$, for $u = 2, 3, \ldots, N + 1$, and $f(u) = (2N + 1 - u)/N^2$, for $u = N + 2, \ldots, 2N$.

Example 3.11   Distribution of $U = X * Y$.   If $X_1, X_2, \ldots, X_n$ are IID DUNI(N) with the same range, find the distribution of $\min(X_1, X_2, \ldots, X_n)$.

Solution: The convolution of $(a_1, a_2, \ldots, a_n)$ and $(b_1, b_2, \ldots, b_n)$ is defined as $(c_k = \sum_{j=1}^{k} a_j b_{k-j})$. As $f(x) = f(y) = 1/N$, $c_1 = 1/N^2$, $c_2 = 1/N^2 + 1/N^2 = 2/N^2$, and so on, $c_k = k/N^2$. As the sum of the probabilities should be unity, the PMF of U is given by $f(u, N) = (2N/(N+1))u/N^2 = 2u/(N(N+1))$, for $u = 1, 2, \ldots, N$.

Example 3.12   Distribution of the minimum for DUNI(N)   If $X_1, X_2, \ldots, X_n$ are IID DUNI(N), find the distribution of $Y = X_{(1)} = \min(X_1, X_2, \ldots, X_n)$.

Solution: Let $S_1(y)$ and $F_1(y)$ denote the SF and CDF of $Y = X_{(1)}$, and $F(x)$ denote the CDF of X. Then $S_1(y) = 1 - F_1(y) = \Pr[Y > y] = \Pr[X_1 > y] \Pr[X_2 > y] \ldots \Pr[X_n > y]$ because of independence. As each of the X_i's are IID, this becomes $S_1(y) = [1 - F(y)]^n$. The PMF of Y is obtained by subtraction $F_1(y) - F_1(y-1)$, which is the same as $S_1(y-1) - S_1(y)$ (if $S_1(y)$ is defined as $\Pr[Y > y]$) and $S_1(y) - S_1(y+1)$ (if $S_1(y)$ is defined as $\Pr[Y \geq y]$). As the CDF of $X \sim$ DUNI(N) is x/N, we get the SF as $1 - x/N$. Substitute in the above to get the PMF as $(1 - (y-1)/N)^n - (1 - y/N)^n$. Take N as a common factor to get the PMF as $f_1(y) = [(N - y + 1)^n - (N - y)^n]/N^n$. This is the required PMF of $Y = X_{(1)}$.

Example 3.13   Distribution of the maximum for DUNI(N)   If $X_1, X_2, \ldots, X_n$ are IID DUNI(N), find the distribution of $Y = X_{(n)} = \max(X_1, X_2, \ldots, X_n)$.

Solution: Let $F_1(y)$ denotes the CDF of $Y = X_{(n)}$. Then $F_1(y) = \Pr[Y \leq y] = \Pr[X_1 \leq y] \Pr[X_2 \leq y] \ldots \Pr[X_n \leq y]$ because of independence. As each of the X_i's are IID, this becomes $F_1(y) = [F(y)]^n$. The PMF of Y is obtained by subtraction $F_1(y) - F_1(y-1)$. As the CDF of $X \sim$ DUNI(N) is x/N, we get $F_1(y) = (y/N)^n - ((y-1)/N)^n$. Take N as a common factor to get the PMF as $f(y) = [y^n - (y-1)^n]/N^n$.

Problem 3.14   Find the variance of a distribution defined as $f(x, n) = 1/(1 + 12\sigma^2)^{1/2}$, for $x = 1, 2, \ldots, n$.

3.4 RELATION TO OTHER DISTRIBUTIONS

For $N = 1$ we get the degenerate distribution, and $N = 2$ results in Bernoulli distribution BER(0.5). Consider a distribution with PMF $f(x, p) = Cp^{F(x)}(1 - p)^{1-F(x)}$ where $F(x)$ is the CDF (or SF) of any discrete distribution, $0 < p < 1$, and C is a constant. This can be written in the alternate form $f(x, p) = C(1 - p)[p/(1 - p)]^{F(x)}$. If $F(x)$ represents the CDF of DUNI(N), we get $f(x, p) = C(1 - p)[p/(1 - p)]^{x/N}$, for $x = 0, 1, 2, \ldots, N - 1$. This is a special case of geometric distribution for $p < 0.5$, discussed in the next chapter. Similarly, the random variable $Y_N = X_N/N$ takes the values $[0, 1/N, 2/N, \ldots, 1]$ and approaches the continuous uniform distribution on unit interval as $N \to \infty$.

3.5 APPLICATIONS

This distribution finds applications in quality control, sampling theory, random number generation, and in many engineering fields. The variate values are non-integers in some engineering applications. Consider $f(x; N) = 1/N$ for $x = a, a + h, a + 2h, \ldots, a + (N - 1)h$. If $Y \sim$ DUNI(N) with domain 0 to $N - 1$, this distribution can be obtained by the linear transformation $X = a + y * h (X = a + (y - 1)h$ if the domain of Y is 1 to N) so that $E(X) = a + hE(Y) = a + h(N - 1)/2$. Using $Var(aX + b) = a^2 Var(X)$, we get $Var(X)$ as $h^2 Var(Y) = h^2(N^2 - 1)/12$. The third moment is obviously zero due to the symmetry. The fourth moment is $h^4(N^2 - 1)(3N^2 - 7)/240$.

3.5.1 SAMPLING

Suppose we have a large number of N labeled items from which a sample of size n is to be taken. We could generate a random sample of size n from a discrete distribution with domain $\{1, 2, \ldots, N\}$, and choose corresponding labeled item as our random sample. Single digit random numbers are assumed to be generated by a DUNI(10) distribution.

3.5.2 RANDOM NUMBERS

The DUNI(N) is used in lotteries and random sampling. Let there be m prizes in a lottery. If N tickets are sold, the chance that an arbitrary ticket will win a prize is m/N. If each ticket is printed with the same number of digits (say the width is 6), and each of the digits $(0, 1, \ldots, 9)$ are equally likely, the PMF of kth digit is DUNI(10). Similarly, in random sampling with replacement, if the population has N elements, the probability distribution of the kth item in the sample is DUNI(N).

3.6   SUMMARY

This chapter introduced the discrete uniform distribution, and described its properties. Distribution of the minimum and maximum of several discrete uniform random variables are discussed. Its relation to other distributions are briefly described. It is extensively used in random number generation and sampling with replacement.

CHAPTER 4

Geometric Distribution

After finishing the chapter, readers will be able to ...

- Understand geometric distributions.

- Explore properties of geometric distribution.

- Discuss arithmetico-geometric distribution.

- Apply geometric distribution to practical problems.

4.1   DERIVATION

Consider a sequence of independent Bernoulli trials with the same probability of success p. We observe the outcome of each trial, and either continues it if it is not a success, or stops it if it is a success. This means that if the first trial results in a success, we stop further trials. If not, we continue observing failures until the first success is observed. Let X denote the number of trials needed to get the first success. Naturally, X is a random variable that can theoretically take any value from 0 to ∞. In summary, practical experiments that result in a geometric distribution can be characterized by the following properties.

1. The experiment consists of a series of IID Bernoulli trials.

2. The trials can be repeated independently without limit (as many times as necessary) under identical conditions. The outcome of one trial has no effect on the outcome of any other, including next trial.

3. The probability of success, p, remains the same from trial to trial until the experiment is over.

4. The random variable X denotes the number of trials needed to get the first success ($q^{x-1}p$), or the number of trials preceding the first success ($q^x p$).

If the probability of success is reasonably high, we expect the number of trials to get the first success to be a small number. This means that if $p = 0.9$, the number of trials needed is much less than if $p = 0.5$, in general. If a success is obtained after getting x failures, the probability is $f(x; p) = q^x p$ by the independence of the trials. This is called the geometric distribution, denoted by $GEO(p)$. It gets its name from the fact that the PMF is a geometric progression (GP) with first term p, and common difference q with closed form $(p/(1-q))$. In other words, the individual probabilities (divided by the first probability p) form a GP. It has a single parameter p, the probability of success in each trial. Sometimes θ is used instead of p in engineering, and ρ is used in queuing theory and stochastic processes.

Although the classical geometric distribution has infinite range, most of the practical applications are observed for a finite range (of x values) only. A finite GP

$$\sum_{k=0}^{n} q^k = \left(1 - q^{n+1}\right)/(1-q) \quad \text{for} \quad q \neq 1, \tag{4.1}$$

can be converted into a *finite* geometric distribution (by cross-multiplication) as:

$$f(x, p) = q^x p/\left(1 - q^{n+1}\right) \quad \text{for} \quad x = 0, 1, \ldots, n. \tag{4.2}$$

The PMF becomes $f(x; p) = pq^{x-1}/(1-q^n)$ when the range is 1 to n. This is the right-truncated geometric distribution discussed below. As n tends to ∞, q^{n+1} tends to zero because q being a probability is < 1. This will then become the classical geometric distribution.

The $GEO(p)$ is used to model the number of complementary events $\overline{E}$ before the first event E in repeated mutually independent Bernoulli trials. Denoting success by S and failure by F, it can be considered as the sequence $S, FS, FFS, FFFS, \ldots$ which is the expansion $(1 - F)^{-1} S = S/(1-F)$ where F^n is written as n F's (e.g., $F^3 = FFF$).

This can be considered as the distribution of waiting time until the occurrence of first success. Consider a sequence of customers in a service queue. Assume that either a new customer joins the queue (with probability p) or none arrives (with probability $q = 1 - p$) in a short-enough time interval. The time T until the next arrival is distributed as $GEO(p)$.

4.1.1 ALTERNATE FORMS

The PMF of a zero-truncated geometric distribution (ZTGD) is $f(x; p) = q^{x-1}p$, where $x = 1, 2, \ldots, \infty$ (see discussion below). We could combine the above two cases and write the PMF as

$$f(x; p) = \begin{cases} q^{x-1}p & \text{if } x \text{ ranges from } 1, 2, 3, \ldots, \infty \\ q^x p & \text{if } x \text{ ranges from } 0, 1, 2, \ldots, \infty \\ 0 & \text{elsewhere.} \end{cases}$$

The second form follows easily by a change of origin transformation $Y = X - 1$ in the first form. The mean, mode and other location measures are simply displaced in this case. The variance remains the same because $V(X - 1) = V(X)$. The Polya distribution

$$f(x; \lambda) = [\lambda/(1 + \lambda)]^x/\lambda, \quad x = 1, 2, \ldots \tag{4.3}$$

is obtained by setting the mean $q/p = \lambda$ so that $1/p = (1 + \lambda)$ or $p = 1/(1 + \lambda)$. Now substitute in $q^{x-1}p$, multiply numerator and denominator by λ to get the above form. Similarly, put $1/p = \lambda$ so that $p = 1/\lambda$ and $q = (\lambda - 1)/\lambda$ to get another form as

$$f(x; \lambda) = [(\lambda - 1)/\lambda]^x/\lambda = (1 - 1/\lambda)^x/\lambda, \quad x = 0, 1, 2, \ldots. \tag{4.4}$$

Set $\log(1/p) = \lambda$ to get

$$f(x; \lambda) = [1 - \exp(-\lambda)]^x/\exp(\lambda), \quad x = 0, 1, 2, \ldots \tag{4.5}$$

and $\log(1/q) = \lambda$ to get

$$f(x; \lambda) = \exp(-\lambda x)(1 - \exp(-\lambda)), \quad x = 0, 1, 2, \ldots. \tag{4.6}$$

Put $q = P/Q$ where $Q = 1 + P$ to get another form $f(x; p) = (P/Q)^x/Q$, for $x = 0, 1, 2, \ldots, \infty$. The PGF is $E(t^x) = \sum_{k=0}^{\infty}(P/Q)^x/Qt^x = (1/Q)[1/(1 - (P/Q)t)] = 1/(Q - Pt)$. This is called Furry distribution in high-energy physics.

Problem 4.1   Find the distribution of the number of rolls of two fair dice until the occurrence of a double six.

Problem 4.2   Pizza Hut puts 10 discount coupons arbitrarily in 100 pizza boxes. Assuming that a customer will immediately claim the offer, find the probability of the following: (i) at least one claim is received in the next four pizza orders, and (ii) a claim is received on the fifth pizza sold.

4.2   RELATION TO OTHER DISTRIBUTIONS

It is a special case of the negative binomial distribution when $r = 1$. If $X_1, X_2, \ldots, X_n$ are IID geometric variates with parameter p, then $Y = X_1 + X_2 + \cdots + X_n$ has a negative binomial distribution with parameters n, p [99]. This is easily proved using the MGF. It can also be considered as the discrete analogue of the exponential distribution. A geometric distribution $GEO(p)$ can be obtained as the distribution of integer part of an exponential distribution $EXP(\lambda)$ where $p = 1 - \exp(-\lambda)$, or equivalently $\lambda = \log(1/q)$. As noted above, we could reparametrize the geometric distribution by putting $\lambda = q/p$ to get

$$f(x; \lambda) = (\lambda/(1 + \lambda))^x/(1 + \lambda) = \lambda^x/(1 + \lambda)^{x+1}, \tag{4.7}$$

with mean λ and variance $\lambda(1 + \lambda)$.

If X has a zero-truncated $GEO(p)$ with PMF $f(x; p) = q^{x-1}p$, then $Y = \log(X)$ has PMF $f(y; p) = (p/q)q^{\exp(y)}$, for $y = 0, \log(2), \log(3), \ldots$. The mean $(X_1 + X_2 + \cdots + X_n)/n$

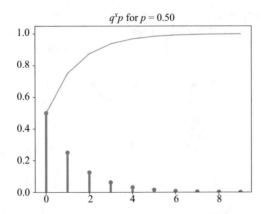

Figure 4.1: PMF and CDF of geometric distribution ($p = 0.50$).

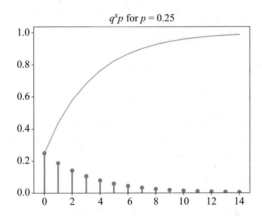

Figure 4.2: Geometric distribution for $p = 0.25$.

tends to the negative binomial distribution when n increases, and will eventually tend to normality as $n \to \infty$ as shown in next chapter. Although the change of origin and scale transformation $Y = a + hX$ can be applied on geometric random variables, they are not popular.

Swapping the roles of p and q results in a dual geometric distribution $\text{DGEO}(p)$. Thus, if Y denotes the number of successes preceding the first failure, we have $f(y, p) = p^y q$, $y = 0, 1, \ldots, \infty$. This finds applications in several manufacturing industries, and engineering fields. This has mean p/q and variance p/q^2. The dual form of (4.2) is $f(x, p) = p^x q/(1 - p^{n+1})$ for $x = 0, 1, \ldots, n$. In the shifted version where X denotes the number of trials needed to get the first failure, we have $f(x, p) = p^{x-1} q$, $x = 1, 2, \ldots, \infty$. These two are related as $X = Y + 1$, so that the mean and variance are $1/q$ and p/q^2. It is a special case of discrete Weibull distribution $\text{DWEIB}(p, \beta)$ when $\beta = 1$.

A geometric distribution of order k is an extension. In a sequence of independent Bernoulli trials, we look for the first occurrence of a consecutive block of k successes (either in the beginning itself, or surrounded by failures). For example, in SSFFSFSSF SSSF, an SSS occurs at position 10. The PGF of a geometric distribution of order k is one in which prior events are either all failures or success runs of length less than k is given by $P(t) = q^k t^k (1 - pt)/[1 - t + qp^k t^{k+1}]$.

4.3 PROPERTIES OF GEOMETRIC DISTRIBUTION

Both the above-mentioned densities are related through a change of origin transformation $Y = X - 1$. This simply displaces the distribution to the left or right. Using $E(Y) = E(X) - 1$, we get $E(Y) = (1/p - 1) = (1 - p)/p = q/p$. Variance remains the same because $V(Y) = V(X)$. As $\sigma^2 = \mu/p > \mu$, the distribution is over-dispersed. The ratio $\Pr[X = 1]/\Pr[X = 0]$ called the incidence jump rate (IJR) is q, and the excess dispersion index (EDI) is $1/p$ (Shanmugam and Radhakrishnan (2011) [93]).

Similarly, $Z = \min(X_1, X_2, \ldots, X_n)$ has the same geometric distribution. The mode is always the first value in the support set (which is 0, if the domain is $(0, 1, \ldots, \infty)$, or 1, if domain is $(1, 2, \ldots, \infty)$).

Example 4.3 Median of geometric distribution Prove that the median of a geometric distribution is $\lfloor -\log(2)/\log(q) \rfloor$ when the domain is $(0, 1, \ldots, \infty)$.

Solution: Let $m - 1$ be the median (note that the very first x value (i.e., $x = 0$) can be the median when $p \geq 1/2$, and that is why it is taken as $m - 1$). Then by definition $p \sum_{x=0}^{m-1} q^x \geq 1/2$. Using the sum of GP this becomes $p[1 - q^m]/(1 - q) \geq 1/2$, which simplifies to $[1 - q^m] \geq 1/2$, or equivalently $q^m < 1/2$. Take log to get $m \log(q) < -\log(2)$, from which $m < -\log(2)/\log(q)$. When $p = 2/3$, we have $q = 1/3$ so that the RHS is $-\log(2)/-\log(3) < 1$. That is why the integer part is taken to get the median. When the range starts with 1, we have $p \sum_{x=1}^{m-1} q^x \geq 1/2$. This becomes $pq[1 + q + q^2 + \cdots + q^{m-1}] \geq 1/2$. This simplifies to $q[1 - q^m] \geq 1/2$. Proceed as above to get $q^m < 1 - 1/(2q)$. Take log to get $m < \log(1 - 1/(2q))/\log(q)$. As $2q - 1$ is $q - p$, this becomes $m < \log(q - p)/\log(q)$. Take integer part of RHS to get the median.

Coefficient of skewness is $\beta_1 = (2 - p)/\sqrt{q} = (1 + q)/\sqrt{q} = \sqrt{q} + 1/\sqrt{q}$. As the numerator never vanishes for valid values of $0 \leq p \leq 1$, the geometric distribution is never symmetric (in fact it is always positively skewed). The kurtosis is $\beta_2 = (p^2 + 6q)/q = 6 + p^2/q$. As p^2/q can never be negative, the distribution is always leptokurtic (see Table 4.1). Moreover, $p^2/q \to \infty$ as $q \to 0$. The PGF is $p/(1 - qt)$ and the ChF is $p/(1 - qe^{it})$.

The PGF and ChF of the zero-truncated GEO(p) are $pt(1 - qt)$ and $p \exp(it)/(1 - q \exp(it))$, respectively. The additivity property can be used to generate random numbers from negative binomial distribution using a random number generator for geometric distribution. Similarly, if X_i are independent $GEO(p_i)$, then $Z = \min(X_1, X_2, \ldots, X_n)$ has the same geometric distribution $GEO(P)$ where $P = 1 - \prod_{i=1}^{n}(1 - p_i)$. This is discussed below.

4.3.1   CDF AND SF

The CDF is obtained as $F(x) = \sum_{k=0}^{x} q^k p = p[1 + q + q^2 + \cdots + q^x]$. As q is a probability, each power of q is between 0 and 1. Hence, the above series converges for all values of q, giving the summed value $p[1 - q^{x+1}]/[1 - q]$. As $(1 - q) = p$, the p in the numerator and denominator cancels out giving the CDF as $F(x) = [1 - q^{x+1}] = [1 - (1 - p)^{x+1}]$. The SF follows from this as $S(x) = \Pr[X > x] = 1 - [1 - (1 - p)^{x+1}] = (1 - p)^{x+1}$. Note that if SF is defined as $\Pr[X \geq x]$ this becomes $q^x = (1 - p)^x$. The hazard function is $f(x; p)/S(x; p) = q^x p/q^{x+1} = p/q$ (and it is p if SF is defined as $\Pr[X \geq x]$). The cumulative hazard function is

$$H(x) = -\ln S(x) = -\ln\left((1 - p)^x\right) \qquad x = 0, 1, 2, \ldots . \qquad (4.8)$$

The inverse distribution function is $F^{-1}(u) = \left\lfloor \frac{\ln(1-u)}{\ln(1-p)} \right\rfloor$ $0 < u < 1$.

4.3.2   TAIL PROBABILITIES

The survival probabilities from $x = c$ is

$$\sum_{x=c}^{\infty} q^x p = p\left[q^c + q^{c+1} + \cdots\right] = pq^c\left[1 + q + q^2 + \cdots\right] = pq^c/(1 - q) = pq^c/p = q^c.$$

As $q < 1$, this goes down to zero for large c. The left-tail probabilities can be found from complementation as $\Pr[0 \leq x \leq c] = 1 - \Pr[x > c] = 1 - q^{c+1}$.

4.3.3   DISTRIBUTION OF EXTREMES

Suppose a random sample of size n is drawn from a $GEO(p)$. As the distribution is leptokurtic and tails off to the right, chances are that there are more data points in the extreme left tail. As shown in Chapter 6, the Poisson distribution also exhibits this property for very small parameter values, where inter-arrival times between two data points is exponentially distributed. The Poisson distribution is equi-dispersed ($\mu = \sigma^2$), while the geometric distribution is over-dispersed ($\mu < \sigma^2$). The following paragraphs derive the distribution of the minimum and maximum among a random sample drawn from a geometric distribution.

Example 4.4   Distribution of the minimum for GEO(p)   If $X_1, X_2, \ldots, X_n$ are IID GEO(p), find the distribution of $Y = X_{(1)} = \min(X_1, X_2, \ldots, X_n)$.

Solution: Let $S_1(y)$ and $F_1(y)$ denote the SF and CDF of $Y = X_{(1)}$, and $F(x)$ denote the CDF of X. Then $S_1(y) = 1 - F_1(y) = \Pr[Y > y] = \Pr[X_1 > y]\Pr[X_2 > y]\ldots\Pr[X_n >$

y] because of independence. As each of the X_i are IID, this becomes $S_1(y) = [1 - F(y)]^n$. The PMF of Y is obtained by subtraction $F_1(y) - F_1(y - 1)$, which is the same as $S_1(y - 1) - S_1(y)$ (if $S_1(y)$ is defined as $\Pr[Y > y]$) and $S_1(y) - S_1(y + 1)$ (if $S_1(y)$ is defined as $\Pr[Y \geq y]$). As the CDF of $X \sim \text{GEO}(p)$ is $[1 - q^{x+1}]$, we get the SF as q^{x+1}. Substitute in the above to get the PMF as $[q^y]^n - [q^{y+1}]^n = q^{ny} - q^{n(y+1)}$. This simplifies to $q^{ny}[1 - q^n]$. Substitute $Q = q^n$ to get the PMF as $Q^y[1 - Q]$, which is $\text{GEO}(P)$ where $P = 1 - Q = 1 - q^n$. If the X_i are independent, but distributed with different probabilities p_i's as $\text{GEO}(p_i)$, $Y = X_{(1)} \sim \text{GEO}(P)$ where $P = 1 - Q = 1 - \prod_{i=1}^n q_i$. This is the required PMF of $Y = X_{(1)}$.

Example 4.5 Distribution of the maximum for GEO(p) If $X_1, X_2, \ldots, X_n$ are IID $\text{GEO}(p)$, find the distribution of $Y = X_{(n)} = \max(X_1, X_2, \ldots, X_n)$.

Solution: Let $F_1(y)$ denote the CDF of $Y = X_{(n)}$, and $F(x)$ denote the CDF of X. Then $F_1(y) = \Pr[Y \leq y] = \Pr[X_1 \leq y] \Pr[X_2 \leq y] \ldots \Pr[X_n \leq y]$ because of independence. As each of the X_i's are IID, this becomes $F_1(y) = [F(y)]^n$. The PMF of Y is obtained by subtraction $F_1(y) - F_1(y - 1)$. As the CDF of $X \sim \text{GEO}(p)$ is $[1 - q^{x+1}]$. Substitute in the above to get the PMF as $(1 - q^{y+1})^n - (1 - q^y)^n$. This is the required PMF of $Y = X_{(n)}$. In the particular case when $n = 2$, this becomes $f(y, p) = (1 - q^{y+1})^2 - (1 - q^y)^2$. Expand as a quadratic and cancel out common factors to get $2q^y(1 - q) - (q^{2y}(1 - q^2))$. Write $1 - q^2 = (1 + q)(1 - q) = p(1 + q)$ to get $f(y, p) = q^y p[2 - (1 + q)q^y]$.

Problem 4.6 If $X_1, X_2, \ldots, X_n$ are IID $\text{GEO}(p)$, find the distribution of $V = \max(X_1, X_2, \ldots, X_n)$ and $U = \min(X_1, X_2, \ldots, X_n)$, and prove that $W = V - U$ and U are independent.

4.3.4 MEMORY-LESS PROPERTY

The geometric density function possesses an interesting property called memory-less property.

Theorem 4.7 *If m and n are natural numbers, and $X \sim \text{GEO}(p)$, then $\Pr(X > m + n | X > m) = \Pr(X > n)$.*

Proof. We know that $P(A|B) = P(A \cap B)/P(B)$. Applying this to the LHS we get $\Pr(X > m + n | X > m) = \Pr(X > m + n \cap X > m)/P(X > m)$. But the numerator is simply $\Pr(X > m + n)$, so that the ratio becomes $\Pr(X > m + n)/P(X > m)$. Substituting the PMF, this becomes $\sum_{x=m+n+1}^{\infty} q^x p / \sum_{x=m+1}^{\infty} q^x p = q^n$ (see below), which is $\Pr(X > n)$. The above result holds even if the $>$ operator is replaced by $\geq$. □

4.3.5   MOMENTS AND GENERATING FUNCTIONS

The mean and variance are $\mu = 1/p$, $\sigma^2 = q/p^2$ if $f(x; p) = q^{x-1}p$, with support $x = 1, 2, \ldots$; and q/p, q/p^2 if $f(y; p) = q^y p$ with support $y = 0, 1, 2, \ldots$. The mean is easily obtained as

$$\mu = \sum_{x=0}^{\infty} x q^x p = p \left[q + 2q^2 + 3q^3 + \cdots\right] = pq[1-q]^{-2} = pq/p^2 = q/p. \qquad (4.9)$$

If the PMF is taken as $f(x; p) = q^{x-1}p$, the mean is $1/p$. If $x_1, x_2, \ldots, x_n$ is a random sample of size n from GEO(p), the MLE is $1/\bar{x} - 1$ in the first case and $1/\bar{x}$ in the second case. If a single observed value (say k) is available, then the MLE of p is $1/k$.

Now consider, $E(X^2) = \sum_{x=0}^{\infty} x^2 q^x p$. To evaluate this, start with the identity $p \sum_{x=0}^{\infty} (1-p)^x = 1$. Write this as

$$\sum_{x=0}^{\infty} (1-p)^x = 1/p. \qquad (4.10)$$

Differentiate w.r.t. p to get $\sum_{x=0}^{\infty} x(1-p)^{x-1}(-1) = -1/p^2$. Multiply throughout by $q = 1 - p$, and write this as $\sum_{x=0}^{\infty} x(1-p)^x = (1-p)/p^2$. Differentiate again to get $\sum_{x=1}^{\infty} x^2 (1-p)^{x-1}(-1) = -2/p^3 + 1/p^2 = -(2-p)/p^3$. Multiply both sides by $p(1-p)$, and write $2 - p$ on the RHS as $1 + (p+q) - p = 1 + q$ to get $p \sum_{x=1}^{\infty} x^2 (1-p)^x = q(1 + q)/p^2$. Now use $\sigma^2 = E(X^2) - [E(X)]^2$ to get the variance as $q(1+q)/p^2 - (q/p)^2 = q/p^2$.

This can also be found as follows: $E(X^2) = \sum_{x=0}^{\infty} x^2 q^x p$. Replace p by $1 - q$ and split it into two series: $E(X^2) = \sum_{x=1}^{\infty} (1-q)x^2 q^x = \sum_{x=1}^{\infty} x^2 q^x - q \sum_{x=1}^{\infty} x^2 q^x$. Expand term by term to get $q - q^2 + 4q^2 - 4q^3 + 9q^3 - 9q^4 + \cdots$. This simplifies to $q + 3q^2 + 5q^3 + 7q^4 + \cdots$. Write this as $q[1 + 2q + 3q^2 + 4q^3 + \cdots] + q^2[1 + 2q + 3q^2 + \cdots] = q/p^2 + q^2/p^2$. Now use $\sigma^2 = E(X^2) - [E(X)]^2$ to get q/p^2. Thus, the population variance is a nonlinear function $((1-p)/p^2)$ for the geometric distribution.

If p is very small, both the mean and variance are large. The variance of a geometric distribution can be increased without limit by letting $p \to 0$. We could reparametrize such distributions appropriately to have this asymptotic behavior at the extreme right-end of the parameter space. For instance, let $\lambda = q/p$ for the geometric distribution, so that $p = 1/(1 + \lambda)$ (Polya's distribution). Then $p \to 0$ is equivalent to $\lambda \to \infty$.

The ordinary moments of this distribution are easy to find using PGF or MGF:

$$M_x(t) = E\left(e^{tx}\right) = \sum_{x=0}^{\infty} e^{tx} q^x p = p \sum_{x=0}^{\infty} \left(qe^t\right)^x = p/\left(1 - qe^t\right) \qquad (4.11)$$

(the MGF for range 1 to ∞ is $pe^t/(1 - qe^t)$). Characteristic function is obtained from the MGF as $\Phi_x(t) = p/(1 - qe^{it})$. Replace $\exp(t)$ by t to get the PGF as $p/(1 - qt)$. Take log of the MGF to get the CGF as $K_x(t) = \log(p) - \log(1 - q \exp(t))$. See Chattamvelli and Shanmugam (2019) for other GFs.

Example 4.8 Moments of geometric distribution   Obtain the moments of GEO(p) distribution using MGF.

Solution: Take logarithm to the base e of the MGF and differentiate once, to obtain

$$M_x'(t)/M_x(t) = qe^t / \left[1 - qe^t\right] = q / \left[e^{-t} - q\right]. \tag{4.12}$$

Putting $t = 0$, we get $M_x'(0)/M_x(0) = q/(1 - q) = q/p$. Hence, $\mu = q/p$, as expected. Differentiating again, we get

$$\left[M_x(t)M_x''(t) - M_x'(t)^2\right]/M_x(t)^2 = q\left[(1 - qe^t)e^t + e^t qe^t\right]/\left(1 - qe^t\right)^2, \tag{4.13}$$

from which $M_x''(0) = q/p^2 + q^2/p^2$, so that the variance is q/p^2.

Example 4.9 Mean deviation of geometric distribution   Find the MD of geometric distribution using theorem in Chapter 1.

Solution: We know that the mean of GEO(p) distribution is $\mu = q/p$. Using theorem in Chapter 1, the MD is given by

$$MD = 2 \sum_{x=ll}^{\mu-1} F(x) = 2 \sum_{x=0}^{c} \left[1 - q^{x+1}\right], \quad \text{where } c = \lfloor q/p \rfloor - 1. \tag{4.14}$$

Split this into two sums. The first one becomes $2(c + 1) = 2\lfloor q/p \rfloor$. The sum

$$\sum_{x=0}^{c} q^{x+1} = q\left[1 + q + q^2 + \cdots + q^c\right] = q\left[1 - q^{c+1}\right]/p = (q/p)\left[1 - q^{\lfloor q/p \rfloor}\right]. \tag{4.15}$$

Combine with the first term to get

$$MD = 2(\lfloor q/p \rfloor) - 2(q/p)\left[1 - q^{\lfloor q/p \rfloor}\right]. \tag{4.16}$$

Write $q = 1 - p$ and cancel out $+2$ and -2. This simplifies to

$$2\lfloor 1/p \rfloor \left(q^{\lfloor q/p \rfloor}\right). \tag{4.17}$$

Write $q = 1 - p$ in the exponent to get an alternate expression

$$MD = (2/p)\lfloor 1/p \rfloor \left(q^{\lfloor 1/p \rfloor}\right). \tag{4.18}$$

Example 4.10 $Y = \lfloor X \rfloor$ of an exponential distribution   If X has an exponential distribution, find the distribution of $Y = \lfloor X \rfloor$.

Solution: As X is continuous, $\Pr[Y = y] = \Pr[y \le X < y + 1]$. Now consider

$$\Pr[y \le X < y + 1] = \int_{y}^{y+1} \lambda \, \exp(-\lambda x) dx = -\exp(-\lambda x)|_{y}^{y+1}$$

$$= \exp(-\lambda y) - \exp(-\lambda (y + 1)) = \exp(-\lambda y)[1 - \exp(-\lambda)]. \qquad (4.19)$$

Write $\exp(-\lambda y)$ as $[\exp(-\lambda)]^y$. Then (4.24) is of the form $q^y p = (1 - q)q^y$ where $q = \exp(-\lambda)$. This is the PMF of a geometric distribution with probability of success $p = 1 - q = [1 - \exp(-\lambda)]$. Hence, $Y = \lfloor X \rfloor$ is GEO$([1 - \exp(-\lambda)])$. An interpretation of this result is that a geometric distribution GEO$(1 - \exp(-\lambda))$ can be obtained by discretizing an exponential distribution with mean $m = 1/\lambda$.

Example 4.11   Geometric probability exceeding $1/p$ If $X \sim$ GEO(p) find the probability that X takes values larger than $1/p$.

Solution: Let $\lfloor 1/p \rfloor$ denote the integer part. Then the required probability is

$$p \sum_{x=\lfloor 1/p \rfloor}^{\infty} q^x = pq^{\lfloor 1/p \rfloor} \left(1 + q + q^2 + \cdots\right) = pq^{\lfloor 1/p \rfloor}(1 - q)^{-1} = q^{\lfloor 1/p \rfloor}. \qquad (4.20)$$

Example 4.12   Factorial moments of geometric distribution   Obtain the factorial moments of GEO(p) distribution.

Solution: Differentiate the identify $\sum_{x=0}^{\infty} q^x = 1/(1 - q)$ w.r.t. q multiple times to obtain the factorial moments. Differentiating it once, we get

$$\sum_{x=0}^{\infty} xq^{x-1} = 1/(1 - q)^2 = 1/p^2. \qquad (4.21)$$

Multiply both sides by pq. Then the LHS becomes $\sum_{x=0}^{\infty} xq^x p = E(X)$. The RHS is $pq/p^2 = q/p$. Differentiating it again results in

$$\sum_{x=1}^{\infty} x(x - 1)q^{x-2} = 2/(1 - q)^3 = 2/p^3. \qquad (4.22)$$

Multiply both sides by $q^2 p$ and simplify to get $E[X(X-1)] = 2q^2/p^2$. Differentiating k times gives factorial moment as

$$\sum_{x=k}^{\infty} x(x-1)\ldots(x-k+1)q^{x-k} = 1,2,3,\ldots k/(1-q)^{k+1}. \tag{4.23}$$

Multiply both sides by $q^k p$ to get

$$E[X(X-1)\ldots(X-k+1)] = k!/(1-q)^{k+1}q^k p = k!q^k/p^k = k!(q/p)^k. \tag{4.24}$$

This gives the recurrence for factorial moments as $\mu_{(k+1)} = (q/p)(k+1)\mu_{(k)}$.

The above results could also be obtained using the factorial MGF (FMGF) $E(1+t)^x = p/[1-q(1+t)] = p/(p-qt)$ (Chattamvelli and Shanmugam (2019), p. 57) [18]. As ordinary powers of x are related to falling factorials as

$$x^k = \sum_{j=1}^{k} \left\{ {k \atop j} \right\} x_{(j)},$$

where $\left\{ {k \atop j} \right\}$ is the Stirling numbers of the second kind, which is valid for $k \geq 1$. It is easy to establish a relation among ordinary moments and factorial moments of a geometric distribution using this result. Multiply both sides by $q^{x-1}p$ and sum from 1 to ∞

$$\sum_{x=1}^{\infty} x^k q^{x-1} p = p \sum_{x=1}^{\infty} \sum_{j=1}^{k} \left\{ {k \atop j} \right\} x_{(j)} q^{x-1},$$

to get $\mu_k = \sum_{j=1}^{k} \left\{ {k \atop j} \right\} \mu_{(j)}$. Now substitute $\mu_{(j)} = j!(q/p)^j$ to get $\mu_k = \sum_{j=1}^{k} \left\{ {k \atop j} \right\} j!(q/p)^j$.

Example 4.13   Inverse moments of geometric distribution   Prove that the first inverse moment of a geometric distribution is $\log(p^{-p/q})$.

Solution: By definition $E(1/(x+1)) = \sum_{x=0}^{\infty}(1/(x+1))q^x p = p\sum_{x=0}^{\infty} q^x/(x+1) = (p/q)(-\log(1-q))$. Write this as $(-p/q)(\log(1-q))$ and use $\log(p^k) = k\log(p)$ to obtain the desired form.

Example 4.14   Variance of geometric distribution   Prove that the ratio of variance to the mean of a geometric distribution is $1/p$. Express the variance as a function of μ and discuss the asymptotic behavior.

Solution: We know that the variance is $q/p^2 = (1-p)/p^2$ and mean is q/p. As $p \to 0$, numerator of variance $\to 1$, and the denominator $\to 0$. The ratio $\sigma^2/\mu = (q/p^2)/(q/p) = 1/p$, which is obviously > 1 as $0 < p < 1$. Thus, the ratio tends to ∞. This has the interpretation that as $p \to 0$, the number of trials needed to get the first success increases without limit. The variance is expressed as a function of the mean as $\sigma^2 = \mu(1 + \mu)$.

Problem 4.15   If $X \sim \text{GEO}(p)$ find the distribution of $Y = 1/X$. Find its mean and variance.

4.4   SPECIAL GEOMETRIC DISTRIBUTIONS

There are many special forms of the geometric distribution. This section presents the most popular among them.

4.4.1   ARITHMETICO-GEOMETRIC DISTRIBUTION

As $(1 - q)^{-2} = 1 + 2q + 3q^2 + 4q^3 + \cdots$, an arithmetico-geometric distribution[1] can be obtained from this for $0 < q < 1$ as

$$f(x; p) = (x + 1)q^x p^2, \quad \text{for } x = 0, 1, 2, \ldots, \infty \tag{4.25}$$

($f(x; p) = xq^{x-1}p^2$ for $x = 1, 2, \ldots, \infty$). This distribution has an interesting property that the first inverse moment is p:

$$E(1/(X + 1)) = \sum_{x=0}^{\infty}(1/(x + 1))\,(x + 1)\,q^x p^2 = p^2 \sum_{x=0}^{\infty} q^x = p^2(1 - q)^{-1} = p. \tag{4.26}$$

Truncated and inflated versions of this distribution can also be defined. For instance, the zero-truncated arithmetico-geometric (ZTAG) distribution has PMF

$$f(x; p) = (x + 1)q^x p^2/(1 - p^2) = (x + 1)q^{x-1}p^2/(1 + p), \quad \text{for } x = 1, 2, \ldots, \infty. \tag{4.27}$$

A right-truncated arithmetico-geometric distribution is obtained from the above as

$$g(x; p, k) = (x + 1)q^x p^2/[1 - kq^k/p - q^k/p^2], \quad \text{for } x = k, k + 1, k + 2, \ldots, \infty. \tag{4.28}$$

Problem 4.16   Find the mean and variance of arithmetico-geometric distribution.

4.4.2   LOG-GEOMETRIC DISTRIBUTIONS

If X is a discrete random variable with support $x \geq 1$ such that $\log(X)$ has a geometric distribution, then X is said to have log-geometric distribution. Consider a distribution with PMF $f(x, p) = Cp^{F(x)}(1 - p)^{1-F(x)}$ where $F(x)$ is the CDF (or SF) of any discrete distribution,

---

[1]Also called size-biased geometric distribution, which is a special case of negative binomial distribution (Chapter 5).

$0 < p < 1$, and C is a constant (Chattamvelli (2012) [15]). This can be written in the alternate form $f(x, p) = C(1 - p)[p/(1 - p)]^{F(x)}$. If $F(x)$ represents the CDF of DUNI(N), namely x/N, we get $f(x, p) = C(1 - p)[p/(1 - p)]^{x/N}$, for $x = 0, 1, 2, \ldots, \infty$. Next consider $S(x) = q^{x+1}$ to be the SF of a geometric distribution. Then, substituting in the above gives $f(x, p) = C(1 - p)[p/(1 - p)]^{S(x)} = KR^{q^{x+1}}$, where $R = p/(1 - p)$. Take log of both sides and denote the PMF of the logarithm by $g()$ to get $g(x, p) = D + Eq^{x+1}$ where D and E are constants. The RHS now is a geometric distribution transformed using change of origin and scale. Then Y is called the log-geometric distribution.

4.4.3 TRUNCATED GEOMETRIC DISTRIBUTIONS

Large social media networks like Facebook, Twitter, LinkedIn, etc., are modeled as graphs with nodes representing entities, and edges representing relation among the entities. The same idea is extended to many other applications like collaboration networks among researchers, gene regulatory network, metabolic and biological network, etc. Large-scale networks are considered as a collection of connected components. These are usually modeled using power-law, Pareto, or log-normal distributions. The zero-truncated geometric (ZTG) distribution is an alternative in these cases that gives better fit.

Left-truncated geometric distribution is obtained by truncating at a positive integer k. The resulting PMF is

$$f(x; p) = q^{x+k} p / \left[1 - \sum_{j=0}^{k-1} q^j p \right]. \tag{4.29}$$

As the denominator simplifies to q^k, the resulting distribution belongs to the same family. The zero-truncation is a special case in which the PMF becomes $f(x; p) = q^x p/[1 - p] = q^{x-1} p$ for $x = 1, 2, 3, \ldots$. A right-truncated geometric distribution is used in some applications like modeling rank frequencies of graphemes (in linguistics). If truncation is at $n + 1$, the left-out probability is $q^{n+1} p + q^{n+2} p + \cdots$. Take $q^{n+1} p$ as common factor to get $q^{n+1} p[1 + q + q^2 + \cdots] = q^{n+1}$. Thus, the PMF becomes $f(x; p) = q^x p/[1 - q^{n+1}]$ for $x = 0, 1, 2, \ldots, n$.

If truncation occurs in the left tail at $x = 0$, and the right tail at $n + 1$, the PMF becomes $f(x; p) = q^x p/(1 - p - q^{n+1}) = q^{x-1} p/(1 - q^n)$ for $x = 1, 2, \ldots, n$. If truncation occurs in the left tail at $x = m - 1$, and the right tail at $x = n + 1$, the PMF becomes $f(x; p) = q^x p/(1 - (1 - q^m) - q^{n+1})$ which after simplification becomes $f(x; p) = q^x p/(q^m - q^{n+1})$ for $x = m, m + 1, \ldots, n$. The zero-truncated PMF of reparametrized geometric distribution is

$$f(x; \lambda) = (\lambda/(1 + \lambda))^x/[(1 + \lambda)(1 - 1/(1 + \lambda))] = \lambda^{x-1}/(1 + \lambda)^x, \text{ for } x = 1, 2, 3, \ldots. \tag{4.30}$$

4.4.4   ZERO-INFLATED GEOMETRIC DISTRIBUTION

The truncated geometric distribution was discussed above. A zero-inflated geometric distribution is defined in terms of a ZTG distribution. These are also called dual-state process count models. Consider the distribution defined as

$$f(x; \theta) = \begin{cases} 1 - \theta & \text{if } x = 0, \\ \theta g(x; p) & \text{if } x > 0, \end{cases}$$

where $g(x; p) = q^x p/(1 - p) = q^{x-1} p$ is the ZTG distribution. A suitable choice of θ can improve statistical fit of the data. The MGF is $1 - \theta + (\theta/q) \, p/(1 - q \exp(t))$. The mean is $(1 - \theta) * 0 + \theta \sum_{x=1}^{\infty} x * q^{x-1} p = \theta p (1 - q)^{-2} = \theta/p$. $E(X^2) = \theta q (1 + q)/p^2$, from which $V(X) = \theta q (1 + q)/p^2 - \theta^2/p^2 = (\theta/p^2)(q(1 + q) - \theta)$. It has been used to model trends in rural-urban migration.

A discussion on intervened geometric distribution (IGD) with an application to cardiovascular studies can be found in [8], an oscillating geometric odds distribution in [83], and a tweaked geometric distribution in [84].

Example 4.17   Moments of geometric distribution $q^{x/2} p$ Find the mean of a distribution defined as

$$f(x; p) = \begin{cases} q^{x/2} p & \text{if } x \text{ ranges from } 0, 2, 4, 6, \ldots, \infty \\ 0 & \text{elsewhere.} \end{cases}$$

Solution: By definition $E(X) = \sum_x x q^{x/2} p = p[2q + 4q^2 + 6q^3 + \cdots]$. Take $2q$ as common factor and simplify using

$$(1 - x)^{-2} = 1 + 2x + 3x^2 + 4x^3 + \cdots \tag{4.31}$$

to get $2pq(1 - q)^{-2} = 2pq/p^2 = 2q/p$.

Example 4.18   Conditional distribution of geometric laws   If X and Y are IID GEO(p), find the conditional distribution of $(X|X + Y = n)$.

Solution: As X and Y are independent,

$$\Pr(X|X + Y = n) = \Pr(X = x) * \Pr[Y = n - x]/\Pr[X + Y = n]. \tag{4.32}$$

We will evaluate the denominator expression first. $X + Y$ takes the value n when $x = k$ and $y = n - k$. Hence, $\Pr[X + Y = n] = \sum_{k=0}^{n} P[X = k]P[Y = n - k]$ (here we have terminated the upper limit at n because Y is positive) $= \sum_{k=0}^{n} q^k pq^{n-k} p = (n + 1)p^2 q^n$. Thus,

$$\Pr(X|X + Y = n) = q^x pq^{n-x} p / [(n + 1)p^2 q^n] = 1/(n + 1), \tag{4.33}$$

Table 4.1: Properties of geometric distribution

Property	Expression	Comments
Range of X	0, 1,...,∞(or 1, 2,...,∞)	Discrete, infinite
Mean	$\mu = q/p$ (or $1/p$)	Need not be integer
Variance	$q/p^2 = \mu/p = \mu(\mu + 1)$	$\Rightarrow \mu > \sigma^2$
Mode	0	1 for range 1, 2, 3,...,∞
Median	$\lfloor -\log(2)/\log(q) \rfloor$	$q > 1/2$
Skewness	$\gamma_1 = (1 + q)/\sqrt{q}$	$= (2 - p)/\sqrt{q}$
Kurtosis	$\beta_2 = 9 + p^2/q$	$= 7 + (q + 1/q)$
CV	$1/\sqrt{q}$	
Mean deviation	$(2/p)\lfloor 1/p \rfloor (q^{\lfloor 1/p \rfloor})$	
$E[X(X-1)...(X-k+1)]$	$k!(q/p)^k$	Diverges if $p \to 0$, $k \to \infty$
CDF	$[1 - q^{x+1}]$	
MGF	$p/(1 - qe^t)$	FMGF $= p/(p - qt)$
PGF	$p/(1 - qt)$	
Recurrence	$f(x;n,p)/f(x - 1;n,p) = q$	
Tail probability	q^{x+1}	

Never symmetric, always leptokurtic.

which is the PMF of a discrete uniform distribution DUNI($(n + 1)$).

Example 4.19 Geometric probabilities   If $X \sim \text{GEO}(p)$, find the following probabilities:
(i) X takes even values, and (ii) X takes odd values.

Solution: As the geometric distribution takes $x = 0, 1, 2, \ldots \infty$ values, both the above probabilities are evaluated as infinite sums. (i) $P[X \text{ is even}] = q^0 p + q^2 p + \cdots = p[1 + q^2 + q^4 + \cdots] = p/(1 - q^2) = 1/(1 + q)$.   (ii) $P[X \text{ is odd}] = q^1 p + q^3 p + \cdots = qp[1 + q^2 + q^4 + \cdots] = qp/(1 - q^2) = q/(1 + q)$, which could also be obtained from (i) because $P[X \text{ is even}] = 1 - P[X \text{ is odd}] = 1 - [1/(1 + q)] = q/(1 + q)$.

Example 4.20 Chance of accident   An insurance company finds that a customer drives 1,000 miles per month on the average. From their claims data, they find the probability for a customer to be involved in an accident is one in twelve thousand. What is the probability that an arbitrary customer chosen at random will be in an accident for the first time after 12,000 miles of travel?

Solution: Take 1,000 miles of driving as one unit. Then the probability of first accident ~ $GEO(p = 1/12000)$. If the first accident occurs in 12,000 miles of travel, we can infer that there were no accidents in 12 time units (one year) for which the probability is given by the $GEO(p)$ as $(1 - 1/12000)^{12}(1/12000) = .00008325$.

4.5   RANDOM SAMPLES

Random samples from statistical distributions are generated using different techniques. The inverse distribution function method is the simplest for geometric distribution because the CDF and SF have simple forms. Random samples from this distribution can be generated using a uniform random number u in $(0, 1)$ by first finding a c such that $1 - q^{c-1} < u < 1 - q^c$. Subtract 1 from each term and change the sign to get $q^c < 1 - u < q^{c-1}$. Now consider $q^c < 1 - u$. As $1 - U$ and U have the same distribution, taking log we get $c * \log(q) < \log(u)$ from which $c < \log(u)/\log(q)$. Similarly, taking log of both sides of $1 - u < q^{c-1}$ we get $(c - 1)\log(q) > \log(1 - u)$ or equivalently $c > 1 + \log(u)/\log(q)$. Combine both the conditions to get $c = \lfloor 1 + \log(u)/\log(q) \rfloor$. This value being an integer is returned as the random variate from the geometric distribution.

4.6   APPLICATIONS

Although the geometric distribution was defined in terms of success and failure, the discussion in Chapter 2 applies. This means that any dichotomous event can be considered, and the meaning of success and failures can be swapped. Thus, instead of observing the occurrence of the very first success, suppose an experimenter observes the occurrence of the very first failure. Then the PMF is $f(x, p) = p^x q$, and all other properties are obtained by swapping p and q. This is called the dual geometric distribution. Consider an example of data packets being sent by connectionless protocols like user datagram protocol (UDP). Assuming that the probability of a successful data transmission is p, it remains the same for all data packets being sent independently. If we observe the number of data packets x transmitted until an error occurs in the transmission, it has a $GEO(p)$ distribution. The geometric distribution occurs in steady-state size distribution of $G/M/1$ queuing systems. It is also used to model the state of gaseous mixtures, in mean-energy modeling of oscillators used in crystallography and other fields, in discrete Markov chains, and in game theory.

4.6.1   MANUFACTURING

Acceptance testing is a routine activity in most manufacturing industries. This is especially important for expensive machinery (like medical equipments, avionics instruments, radars, heavy engineering machinery, etc.). Sometimes a rework on finished products can make it acceptable, in case it fails acceptance tests. But semiconductor industry uses defect patterns on wafers to decide whether a chip is acceptable or not. Some of these industries use "killer defects" to decide whether a chip (or instrument) should be discarded or not. Denote the probability of getting a killer defect chip by p. Then the number of acceptable chips produced before a chip is discarded is $GEO(p)$. The negative binomial distribution discussed in the next chapter is better suited to model this purpose.

4.6.2   MACHINE FAILURES

Suppose we count the number of days X between successive failures of an expensive machine. Here we swap the roles of p and q, so that p denotes the probability of failure. Then X is distributed according to $GEO(p)$. The same reasoning applies to mission critical software failures, infrastructure failures, etc.

4.6.3   MEDICAL IMAGING

Modern medical technology uses multiple imaging techniques. The most popular among them are x-rays, γ-rays, computed tomography, ultrasound, and magnetic resonance imaging (MRI). X-rays still remain the most popular choice in under-developed and developing countries due to cost reasons. Some x-rays move in a straight line from source to the detector. These are called primary rays. An x-ray source is assumed to transmit N photons at a specified energy level to a specified point on the detector or image receptor (called a pixel) that records M photons. Both M and N are assumed to be Poisson distributed. Chances are that some photons will pass unaffected through the object and fall on the detector in which case M and N are identical. Other photons that pass through tissues (bones, muscles, etc., in medical x-rays) produces a faithful shadowgram as they are the primary rays that produce useful information for practitioners. Some x-rays undergo Compton scattering and produces secondary x-rays and may reach the recorder. If P and S denote respectively the primary radiation intensity and scattered radiation intensity, the scatter radiation factor is given by $X = S/P$ has a geometric distribution.

4.6.4   TRANSPORTATION

The geometric distribution has also been applied in transportation industry. As a simple example, consider long-distance flights. A flight may reach the destination a few minutes early if there is high-wind blows parallel to the flight path. Flights are delayed if winds are in other directions or due to inclement weather. Suppose a threshold $(-m, +n)$ is setup as an acceptable margin for the flight arrival. If the probability p of reaching the destination within the acceptable margin is known from prior flight data, we could model either a single route (say London–New York), a single flight on multiple days, or all long distance flights to a destination using the geometric distribution.

4.7   SUMMARY

This chapter introduced the basic concepts in geometric distribution. Most important properties of this distribution are derived. Its connection to other distributions are briefly discussed, including zero-truncated and zero-inflated geometric distributions The chapter ends with some applications in various fields.

CHAPTER 5

Negative Binomial Distribution

After finishing the chapter, readers will be able to . . .

- Understand negative binomial distribution and its forms.

- Describe negative binomial distribution and its properties.

- Apply the distribution in practical situations.

- Explore negative binomial regression.

5.1   DERIVATION

The negative binomial distribution is a two-parameter discrete distribution defined on Bernoulli trials. It has a long history, dating back to Blaise Pascal (1679), who discussed it for integer parameter, and Montfort, P.R. (1713), who presented the general form of the PMF. It is known by different names like Pascal distribution (when k is an integer), Polya–Eggenberger distribution in ecology, binomial-waiting time distribution in queuing theory, etc. Practical experiments that result in a negative binomial distribution can be characterized by the following properties.

1. The experiment consists of a series of IID Bernoulli trials.

2. The trials can be repeated independently without limit (as many times as necessary) under identical conditions. The outcome of one trial has no effect on the outcome of any other, including the next trial.

3. The probability of success, p remains the same from trial to trial until the experiment is over.

4. The random variable X denotes the number of trials (or the number of failures) needed to get kth success where k is an integer > 1.

    Consider a sequence of independent Bernoulli trials. Instead of counting the number of trials needed to get the first success, we count the number of failures observed to get the kth success, where k is a fixed integer greater than 1 known in advance[1] (we are actually counting the number of failures, since the number of successes is fixed at k). Every possible outcome of $(k + x)$ trials with k successes and x failures occurs with probability $q^x p^k$. Hence in $x + k - 1$ trials we have observed $k - 1$ successes, and the $(x + k)$th trial must result in the kth success. These two events are independent because the outcomes of $(x + k - 1)$ trials do not affect the outcome at $(x + k)$th trial. The probability of occurrence is thus

$$f(x; k, p) = \binom{x + k - 1}{k - 1} p^{k-1} q^x \times p = \binom{x + k - 1}{k - 1} p^k q^x, \tag{5.1}$$

where $0 < p < 1, k > 0$ and $x = 0, 1, \ldots, \infty$. This is the negative binomial distribution, and is denoted as NBINO(k, p).

5.1.1   ALTERNATE FORMS

Several alternate forms exist for this distribution. Using $\binom{n}{x} = \binom{n}{n-x}$ the PMF becomes

$$f(x; k, p) = \binom{x + k - 1}{x} p^k q^x = \Gamma(x + k)/[\Gamma(k) x!] \, p^k q^x, \tag{5.2}$$

for $x = 0, 1, 2, \ldots$, and $k = 1, 2, \ldots$. For $k = 1$, this reduces to the geometric distribution because $\binom{x}{0} = \binom{x}{x} = 1$. This distribution gets its name from the fact that the successive probabilities are obtained from the infinite series expansion of the expression $p^k (1 - q)^{-k}$, where $q = 1 - p$, and $p > 0$. It differs from BINO(n, p) in that the number of trials is not fixed. The second form in (5.2) is more general, as k is not restricted to be an integer. Next, define the variable Y as the number of trials needed to get the kth success. In $y - 1$ trials we have observed $k - 1$ successes, and yth trial results in a success. This gives an alternate form

$$f(y; k, p) = \binom{y - 1}{k - 1} p^k q^{y+k} \quad \text{for} \quad y = k, \; k + 1, \ldots. \tag{5.3}$$

Here, $y = k$ denotes that the very first k trials are successes. Putting $p = k/(\mu + k)$ and $q = \mu/(\mu + k)$, this could also be written in alternate form as

$$\binom{x + k - 1}{x} (k/(\mu + k))^k (\mu/(\mu + k))^x = \Gamma(x + k)/[\Gamma(k) x!] \, (k/(\mu + k))^k (\mu/(\mu + k))^x.$$

Write $(\mu/(\mu + k))^x$ as $(1 + k/\mu)^{-x}$ and introduce gamma function in place of factorial to get the alternate form

$$f(x; k, \mu) = \Gamma(x + k)/[\Gamma(k) x!] \, (k/(\mu + k))^k (1 + k/\mu)^{-x}.$$

---

[1] k is called the stopping parameter of this distribution in engineering, and as nuisance parameter in epidemiology. This process is called inverse Bernoulli sampling.

Then $p \to 0$ is equivalent to $\mu \to \infty$. These distributions have a characteristic property that the variance is greater than the mean. Naturally, we expect this property to hold in samples drawn from such populations. Such samples are called over-dispersion samples ($s^2 \geq \bar{x}$).

Now set $q = P/Q$ so that $p = 1 - P/Q$ to get the form $f(x; k, P) =$

$$\binom{x+k-1}{x}(1 - P/Q)^k (P/Q)^x = \Gamma(x+k)/[\Gamma(k)\,x!]\,(1/(1+P))^k(P/(1+P))^x. \quad (5.4)$$

This corresponds to the infinite series expansion of $(Q - P)^{-k}$ where $Q = 1 + P$. Now consider

$$f(x; k, p) = \binom{x+k-1}{x} p^k q^x = \Gamma(x+k)/[\Gamma(k)\,x!]\,p^k q^x. \quad (5.5)$$

Swapping the roles of p and q results in a dual negative binomial distribution (DHGD). Instead of counting x items of the second kind, we could count x items of the first kind and k items of the second kind (defectives are assumed as "success" in manufacturing and SQC environments). This gives the PMF $f(x; k, p) = \binom{x+k-1}{x} q^k p^x = \Gamma(x+k)/[\Gamma(k)\,x!]\,q^k p^x$.

Problem 5.1 A haemocytometer has 64 compartments. The probability of finding 10 or more RBC corpuscles in a compartment is 10/64. If the compartments are inspected one by one, what is the probability of finding 10 or more RBC corpuscles in the fifth inspected compartment?

5.2 RELATION TO OTHER DISTRIBUTIONS

Tail areas of binomial and negative binomial distributions are related. That is, $\Pr[Y \geq n - c] = \Pr[X \leq c]$ where $X \sim \text{BINO}(n, p)$ and $Y \sim \text{NBINO}(c, p)$. This can be proved using the following argument. The x in negative binomial distribution denotes the number of failures before the kth success. Consider $Y \sim \text{NBINO}(k, p)$. Then the event $[Y > c]$ means that there are at least $c + 1$ failures preceding the kth success. Hence, $\Pr[Y > c] = 1 - \Pr[Y \leq c]$. The complementary event $[Y \leq c]$ means that there are less than or equal to c failures. Thus, in $c + k - 1$ trials, there exist "c failures and $k - 1$ successes." For fixed c, this probability can be found using binomial distribution as $f(x; k + c - 1, p) = \binom{k+c-1}{k-1} p^{k-1} q^c$. Now we turn our attention only on binomial distribution (X). Vary x from 0 to c to get the probability of at most c failures as $\sum_{j=0}^{c} f(x; k + c - 1, p) = \sum_{j=0}^{c} \binom{k+x-1}{k-1} p^{k-1} q^x$ which is the CDF of $\text{BINO}(k + c - 1, q)$. But from Chapter 2, we know that the sum of left-tail probabilities of $\text{BINO}(n, p)$ is the same as the sum of right-tail probabilities of $\text{BINO}(n, q)$. Hence, the above is the SF of $\text{BINO}(k + c - 1, p)$.

As $k \to \infty$ and $p \to 1$ such that $k(1 - p)$ is a constant, the negative binomial distribution approaches a Poisson law with parameter $\lambda = k(1 - p)$. Symbolically, $\binom{x+k-1}{x} p^k q^x \to \exp(-qk)(qk)^x/x!$; and $\exp(-pk)(pk)^x/x!$ when $p \to 0$. Similarly, $\text{NBINO}(k, p/k)$ as $k \to$

∞ tends to the Poisson law $\exp(-\lambda)\lambda^x/x!$ with complexity $O(kq^2/2)$. This means that "the negative binomial distribution tends to the Poisson law when $p \to 1$ faster than $k \to \infty$." If k is an integer or a half-integer, the SF can be written as $\Pr[Y > y] = I_{1-p}(y,k)$ where $I_x(a,b)$ denotes the incomplete beta function. This can also be written in terms of an F distribution as $\Pr[Y > y] = F_t(2k, 2y)$ where $t = p * y/(q * k)$.

The negative binomial distribution can be regarded as a sum of k independent geometric distributions with the same parameter p [99]. We know from Chapter 4 that the MGF of a GEO(p) distribution is given by $M_x(t) = p/(1 - qe^t)$. Hence, the MGF of k IID GEO(p) is given by $M_Y(t) = [p/(1 - qe^t)]^k$. In general, an NBINO($k, p$) can be decomposed in $\phi(k)$ different ways into sums where each term is negative binomial (or geometric) distributions, where $\phi()$ denotes Euler's partition number (see also Chapter 8, page 161). It is related to the logarithmic distribution as follows: Suppose $Y = X_1 + X_2 + \cdots + X_N$ is a sum of N IID logarithmic series random variables where N itself is Poisson distributed. Then the unconditional distribution of Y is negative binomial (see Chapter 9). Similarly, the zero-truncated NBINO(k, p) tends to the logarithmic distribution as $k \to 0$. It arises as a beta mixture of binomial distributions as well [10]. When the parameter p of a negative binomial distribution has a Beta type-1 distribution, the resulting mixture distribution is a generalized Warring distribution. Consider the PMF

$$f(x; k, p) = \binom{x + k - 1}{x} p^k q^x, \tag{5.6}$$

where p is distributed as $f(p) = [1/B(a,b)]p^{a-1}(1 - p)^{b-1}$. The unconditional distribution is obtained by integrating out p as

$$f(x; k, a, b) = \int_{p=0}^{1} \binom{x + k - 1}{x} p^k q^x [1/B(a,b)]p^{a-1}(1 - p)^{b-1} dp. \tag{5.7}$$

Take constants outside the integral to get

$$f(x; k, a, b) = [1/B(a,b)]\binom{x + k - 1}{x} \int_{p=0}^{1} p^{k+a-1}(1 - p)^{x+b-1} dp. \tag{5.8}$$

Using complete beta integral this becomes $[B(a + k, b + x)/B(a,b)]\binom{x+k-1}{x}$, which is the generalized Warring distribution.

As shown below, the CDF of NBINO(k, p) can be expressed as incomplete beta integral. Such a relationship also exists between the SF of a binomial distribution. This implies that the CDF of NBINO(k, p) can be expressed as SF of BINO(n, p). Thus, $\Pr[X \leq r] = I_p(k, r + 1) = \Pr[Y \geq k]$ where $X \sim$ NBINO(k, p) and $Y \sim$ BINO($k + r, p$). Similarly,

$\Pr[X \leq r] = I_q(r + 1, k) = \Pr[Y \leq k]$. It can also be obtained from the Poisson distribution when the parameter λ has a Gamma distribution.

Example 5.2   Gamma mixture of the Poisson parameter   Prove that a gamma(m, p) mixture of Poisson parameter (λ) gives rise to NBINO($p, m/(m + 1)$) distribution.

Solution: The PMF of Poisson and Gamma variates are, respectively,

$$f(x; \lambda) = e^{-\lambda} \lambda^x / x!, \quad \text{and} \quad g(\lambda; m, p) = \frac{m^p}{\Gamma(p)} e^{-m\lambda} \lambda^{p-1}. \tag{5.9}$$

The unconditional distribution is obtained as $f(x; p) =$

$$\int_{\lambda=0}^{\infty} e^{-\lambda} \lambda^x / x! \frac{m^p}{\Gamma(p)} e^{-m\lambda} \lambda^{p-1} d\lambda$$

$$= \frac{m^p}{x!\Gamma(p)} \int_{\lambda=0}^{\infty} e^{-\lambda(1+m)} \lambda^{x+p-1} d\lambda = \frac{m^p}{x!\Gamma(p)} \frac{\Gamma(p + x)}{(m + 1)^{p+x}}. \tag{5.10}$$

This upon rearrangement becomes

$$f(x; p) = \frac{\Gamma(p + x)}{x!\Gamma(p)} \left(\frac{m}{m + 1}\right)^p \left(\frac{1}{m + 1}\right)^x, \quad \text{for} \quad x = 0, 1, 2, \dots. \tag{5.11}$$

5.3   PROPERTIES OF NBINO(k, p)

An algorithm for the PMF could be obtained from the recurrence $f(x + 1; k, p)/f(x; k, p) = q\,(x + k)/(x + 1)$. By using $\binom{-n}{x} = (-1)^x \binom{n+x-1}{x}$, the PMF can be written alternatively as

$$f(x; k, p) = \binom{-k}{x} p^k (-q)^x = (-1)^x \binom{-k}{x} p^k q^x. \tag{5.12}$$

This distribution is always unimodal and positively skewed.

5.3.1   MGF OF NEGATIVE BINOMIAL DISTRIBUTION

The MGF of NBINO(k, p) is $M_x(t) = E[\exp tx] =$

$$\sum_{x=0}^{\infty} e^{tx} \binom{-k}{x} p^k (-q)^x = \sum_{x=0}^{\infty} \binom{-k}{x} p^k (-qe^t)^x = [p/(1 - qe^t)]^k. \tag{5.13}$$

The PGF is obtained from this as $P_x(r, p; t) = (p/(1 - qt))^r$. The PGF for the shifted version in Equation (5.3) is $t^r (p/(1 - qt))^r$. The CGF is

$$K_x(t) = \log(M_x(t)) = k \log[p/(1 - qe^t)] = k \log(p) - k \log(1 - qe^t). \tag{5.14}$$

The CGF is much more easier to work with because they are location invariant[2] (except for the first cumulant) as $K_1(X + c) = K_1(X) + c$ and $K_n X + c = K_n(X)$ where c is a constant and $K_n(X)$ denotes nth cumulant. More generally, $K_{aX+c}(t) = K_X(at) + ct$. A simpler form for the PGF is obtained for the alternate representation.

5.3.2   MOMENTS OF NEGATIVE BINOMIAL DISTRIBUTION

Differentiate the MGF w.r.t. t and put $t = 0$ to get $E(X) = kq/p$. Alternately, $E(X) = \sum_{x=0}^{\infty} x\binom{x+k-1}{x} p^k q^x$. Expand $\binom{x+k-1}{x}$ and cancel out one x to get $E(X) = \sum_{x=1}^{\infty}(x + k - 1)!/[(x - 1)!(k - 1)!]p^k q^x$. Multiply numerator and denominator by k and take p^k outside the summation to get $E(X) = kqp^k \sum_{x=1}^{\infty}(x + k - 1)!/[(x - 1)!k!]q^{x-1}$. The summation is the binomial expansion of $(1 - q)^{-(k+1)}$. As $(1 - q) = p$, $E(X) = kqp^k/p^{k+1} = kq/p$.

The first inverse moment is

$$E(1/(X + 1)) = \sum_{x=0}^{\infty} 1/(x + 1)\binom{x + k - 1}{x} p^k q^x \tag{5.15}$$

$$= p^k/[q(k - 1)] \sum_{x=0}^{\infty}(x + k - 1)!/[(x + 1)!(k - 2)!]q^{x+1}. \tag{5.16}$$

As the infinite series is $(1 - q)^{-(k-1)} - 1$, we get

$$E(1/(X + 1)) = \left(p^k/[q(k - 1)]\right)\left[(1 - q)^{-(k-1)} - 1\right] = (p/[q(k - 1)])\left(1 - p^{k-1}\right). \tag{5.17}$$

This simplifies to p for $k = 2$, and $p(1 + p)/2$ for $k = 3$.

Example 5.3 Candidate interviews   A company requires k candidates with a rare skill set. As there is a scarcity of local candidates perfectly matching the required skill rset, the company decides to conduct a walk-in interview until all k candidates have been found. If the probability of a candidate who matches perfectly is p, find the expected number of candidates interviewed, assuming that several candidates whose skill set is not completely matching also walks-in.

Solution: We are given that the probability of perfect match is p. An interviewed candidate is either rejected if the skill set is not a 100% match, or hired. As the company needs k such candidates, the distribution of finding all k candidates is negative binomial with parameters (k, p). The expected number of candidates is kq/p. Due to the rarity of the sought skill set, p is small so that q/p is large. For instance, if $p = 0.1$, $q/p = 9$, and if $p = 0.005$, $q/p = 199$.

---

[2]Also called translation invariant.

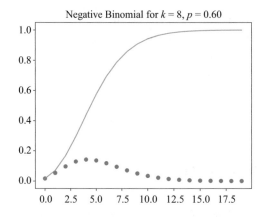

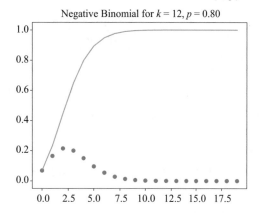

Figure 5.1: PMF and CDF of negative binomial distribution for ($k = 8$, $p = 0.60$).

Figure 5.2: PMF and CDF of negative binomial distribution for ($k = 15$, $p = 0.80$).

5.3.3   VARIANCE GENERATING FUNCTION

A GF can be developed for the variance of a NBINO(k, p) distribution as follows. Consider
$V(t; k, p) = q/[p^2(1 - t)^2]$. Expand as an infinite series to get

$$V(t; k, p) = q/p^2(1 + 2t + 3t^2 + \cdots + kt^{k-1} + \cdots). \tag{5.18}$$

This shows that the coefficient of t^{k-1} in $q/[p^2(1-t)^2]$ is kq/p^2. Thus, the VGF of
NBINO(k, p) is $q/[p^2(1 - t)^2]$. See Chattamvelli and Shanmugam (2019) [18] for other GFs.

Example 5.4   Variance of negative binomial   Prove that the ratio of variance to the mean of
a negative binomial random variable is $1/p$. Discuss asymptotics.

Solution: We know that the variance is given by $V(k, p) = kq/p^2$. Obviously this can be
increased by increasing the parameter k without limit. As the mean is kq/p, the ratio of variance
to the mean is $(kq/p^2)/(kq/p) = 1/p$, which is obviously greater than 1 (as $0 < p < 1$) (over-
dispersed). This gives $p\,\sigma^2 = \mu$, so that $\sigma^2 \to \infty$ as $p \to 0$.

Example 5.5   MD of negative binomial distribution   Find the MD of negative binomial distribution using theorem in Chapter 1.

Solution: We know that the mean of negative binomial distribution is $\mu = kq/p$. Using theorem in Chapter 1, the MD is given by

$$\text{MD} = 2 \sum_{x=ll}^{\mu-1} F(x) = 2 \sum_{x=0}^{c} I_p(k, x+1), \quad \text{where} \quad c = \lfloor kq/p \rfloor - 1, \tag{5.19}$$

where $I(a, b)$ is the incomplete beta function. This simplifies to $2c\binom{k+c-1}{c}q^c p^{k-1} = 2\mu_2 * f_c$ where f_c is the probability mass evaluated at the integer part of the mean. When kq/p is non-integer, a correction term $2\delta F(\lfloor kq/p \rfloor)$ must be added where $\delta = kq/p - \lfloor kq/p \rfloor$.

5.3.4   FACTORIAL MOMENTS

The falling factorial moments are easier to find than ordinary moments. Let $\mu_{(r)}$ denote the rth factorial moment.

Theorem 5.6   *The factorial moment $\mu_{(r)}$ is given by $\mu_{(r)} = k^{(r)} (q/p)^r$.*

Proof. Consider $\mu_{(r)} = E[x_{(r)}] = E[x(x-1)\ldots(x-r+1)]$. Substitute the PMF and sum over the proper range of x to get the RHS as $\sum_x x(x-1)\ldots(x-r+1)\binom{k+x-1}{x}p^k q^x$. Write $x!$ in the denominator as $x(x-1)\ldots(x-r+1) * (x-r)!$, multiply both numerator and denominator by $k(k+1)(k+2)\ldots(k+r-1)$ this becomes

$$k(k+1)(k+2)\ldots(k+r-1) \, p^k q^r \sum_{x=r}^{\infty} \binom{k+x-1}{x-r} q^{x-r}. \tag{5.20}$$

Put $y = x - r$ in (5.14) and rearrange the indexvar. This gives

$$k(k+1)(k+2)\ldots(k+r-1) \, p^k q^r \sum_{y=0}^{\infty} \binom{k+r+y-1}{y} q^y. \tag{5.21}$$

The infinite sum in (5.16) is easily seen to be $(1-q)^{-(k+r)}$. As $(1-q) = p$, the p^k cancels out giving $\mu_{(r)} = E[x_{(r)}] = k^{(r)}(q/p)^r$. This can be written in terms of gamma function as $\mu_{(r)} = [\Gamma(k+r)/\Gamma(k)] \, (q/p)^r$.   □

Factorial Moment Generating Function

The PGF and factorial MGF (FMGF) are related as FMGF(t) = PGF$(1+t)$. This means that if we replace every occurrence of t by $(1+t)$ in the PGF, we get FMGF (Chattamvelli and Shanmugam (2019), pp. 56–58) [18]. As the PGF is $[p/(1-qt)]^k$, we get the FMGF as $[p/(1-q(1+t))]^k = [p/(p-qt)]^k$. Divide numerator and denominator by p^k to get the alternate form FMGF$(t) = (1-qt/p)^{-k}$. Take log to get the factorial cumulant GF (FCGF) as FKGF$(t) = -k\log(1-qt/p)$. Expand as an infinite series using binomial theorem for negative exponent to get

$$(1-qt/p)^{-k} = \sum_{j=0}^{\infty} \binom{k+j-1}{j}(qt/p)^j \tag{5.22}$$

from which the coefficient of t^j is $\binom{k+j-1}{j}(q/p)^j$. Put $j = 1$ and 2 to get the first two values as $\mu_{(1)} = \mu = kq/p$, and $\mu_{(2)} = \binom{k+1}{2}(q/p)^2 = k(k+1)/2(kq/p)^2$, (see Chattamvelli and Shanmugam (2019), p. 15) [18].

Example 5.7 Geometric mixture of a negative binomial parameter Prove that if the parameter k of NBINO(k, p) has a geometric distribution GEO(p_1) with support $[1, 2, \ldots, \infty]$, the resulting mixture is a GEO($pp_1/(1 - pq_1)$) distribution. Find its mean and variance.

Solution: The PMF of $X = $ NBINO(k, p) is $f(x; k, p) = \binom{x+k-1}{x}p^k q^x$, and that of $Y = $ GEO(p_1) is $g(y; p_1) = q_1^{y-1} p_1$, since the support is $[1, 2, \ldots, \infty]$. The unconditional distribution is obtained as

$$f(x, p, p_1) = \sum_{y=1}^{\infty} \binom{x+y-1}{x} p^y q^x q_1^{y-1} p_1. \tag{5.23}$$

Take $p_1 q^x / q_1$ outside the summation to obtain

$$f(x, p, p_1) = (p_1 q^x / q_1) \sum_{y=1}^{\infty} \binom{x+y-1}{x} (pq_1)^y \tag{5.24}$$
$$= (p_1 q^x / q_1) \, pq_1 [1 - pq_1]^{-(x+1)}$$

because the summation reduces to $pq_1[1 - pq_1]^{-(x+1)}$. Consider $1 - q/(1 - pq_1) = (1 - pq_1 - q)/(1 - pq_1) = p(1 - q_1)/(1 - pq_1) = pp_1/(1 - pq_1)$, where we have used $1 - q = p$ and $1 - q_1 = p_1$. Cancel out common factor q_1 and write $[1 - pq_1]^{-(x+1)}$ as $[1 - pq_1]^{-x}/(1 - pq_1)$. Then (5.24) becomes

$$f(x; p, p_1) = [pp_1/(1 - pq_1)] \, (q/(1 - pq_1))^x, \quad \text{for} \quad x = 0, 1, 2, \ldots \tag{5.25}$$

which is GEO($pp_1/(1 - pq_1)$). This has mean $\mu = q/(pp_1)$, which can be obtained directly from the above density by writing $P = pp_1/(1 - pq_1)$, so that $Q = 1 - P = q/(1 - pq_1)$, and the mean is Q/P. Alternately, we could use the formula for the mean of conditional distributions. As $X \sim$ NBINO(R, p) where $R \sim$ GEO(p_1), we have $E(X) = E(E(X|R))$. Now $E(X|R) = Rq/p$. Take expectation of both sides to get $E(X) = E(Rq/p) = (q/p)E(R)$. As R is a zero-truncated geometric distribution with support $[1, 2, \ldots, \infty]$, its mean is $1/p_1$, and variance is q_1/p_1^2. Substitute in the above to get the mean μ as $(q/p)/p_1 = q/(pp_1)$. The conditional variance is $V(X|R) = Rq/p^2$. Now use $V(X) = E[V(X|R)] + V[E(X|R)]$. As $V(X|R) = Rq/p^2$, we have $E[V(X|R)] = E[Rq/p^2] = (q/p^2)E[R] = (q/p^2)(1/p_1) = q/(p_1 p^2)$. As $E(X|R) = Rq/p$, we get $V[E(X|R)] = $

$V(Rq/p) = (q/p)^2 V(R) = (q/p)^2 q_1/p_1^2$. Substitute the values to get the variance as $\sigma^2 = q/(pp_1)^2$.

Problem 5.8   If $X_1, X_2, \ldots, X_N$ are IID exponential random variables where $N \sim$ NBINO(k, p), find the MGF of $S = X_1 + X_2 + \cdots + X_N$.

Problem 5.9   If $X_1, X_2, \ldots, X_N$ are IID logarithmic random variables where $N \sim$ POIS(λ), find the MGF of $S = X_1 + X_2 + \cdots + X_N$.

Problem 5.10   If $X \sim$ NBINO(k, p) and $Y \sim$ GEO(p) (with same p) are IID, find distributions of $U = X + Y$ and $V = X - Y$.

5.4   POISSON APPROXIMATION

As a first try, consider the mean and variance of negative binomial and Poisson distributions. The mean μ is kq/p and variance σ^2 is kq/p^2. Convergence to the Poisson law occurs when $kq/p \simeq kq/p^2$. Cancel out kq/p from both sides to get $1 \simeq 1/p$. This holds when $p \Rightarrow 1$. This implies that the negative binomial distribution tends to the Poisson law when the probability of success p tends to one.

5.4.1   DERIVATION USING PGF

This section derives the exact asymptotic distribution analytically. The PGF of NBINO(k, p) is

$$P_x(k, p; t) = (p/(1 - qt))^k. \tag{5.26}$$

Take log on both sides to get

$$\log(P_x(k, p; t)) = k \log(p/(1 - qt)) = k[\log(p) - \log(1 - qt)]. \tag{5.27}$$

Expand both the logarithmic terms inside square bracket and simplify to get

$$\log(P_x(k, p; t)) = k[(1 - p)(t - 1)]. \tag{5.28}$$

Write $q = 1 - p$, $kq = \lambda$ and exponentiate both sides to get the RHS as $\exp(\lambda(t - 1))$ which is the PGF of Poisson distribution. This proves that the NBINO(k, p) tends to the Poisson law when the probability of success p tends to one. As done in the case of binomial distribution, we could improve upon the approximation by applying (multiplying by) the correction factor $\exp(\lambda + k \log(q))$ when $\lambda < 1$ and probabilities are sought towards the middle (near the mean).

Example 5.11   Mean and variance for the NBINO(k, p)   Find the value of p for which an NBINO(k, p) distribution has variance twice the mean.

Solution: We know that the mean is kq/p and variance is kq/p^2. As the variance is twice the mean, we have $kq/p = kq/p^2$, from which $p = 1/2$. In general, if $p = 1/n$, then the variance is n times the mean. The variance of negative binomial distribution can be increased without limit by letting $p \to 0$.

5.5   MOMENT RECURRENCE

The central moments satisfy the recurrence

$$\mu_{r+1} = q\left((kr/p^2)\,\mu_{r-1} - \partial\mu_r/\partial p\right),\tag{5.29}$$

where $\mu_r = E(x - kq/p)^r$. Consider

$$\mu_r = \sum_{x=0}^{\infty}(x - kq/p)^r\binom{k+x-1}{x}p^k q^x.\tag{5.30}$$

As $\binom{k+x-1}{x}$ is independent of p, and $q = 1 - p$, write the above as

$$\mu_r = \sum_{x=0}^{\infty}\binom{k+x-1}{x}\left\{(x+k-k/p)^r p^k(1-p)^x\right\}.\tag{5.31}$$

Differentiate the expression within the curly brackets w.r.t. p using the function of a function rule to get

$$\partial\mu_r/\partial p = \sum_{x=0}^{\infty}\binom{k+x-1}{x}\left\{r(x+k-k/p)^{r-1}p^k(1-p)^x\left(+k/p^2\right)\right.$$
$$\left. + (x+k-k/p)^r kp^{k-1}(1-p)^x - (x+k-k/p)^r p^k x(1-p)^{x-1}\right\}.\tag{5.32}$$

Combine the last two terms as $[k(1-p) - px] = -p(x - k(1-p)/p) = -p(x+k-k/p)$ to get $-p\,(x+k-k/p)^{r+1}p^{k-1}(1-p)^{x-1}$. Multiply and divide by pq and combine the terms as $-(1/q)\,(x+k-k/p)^{r+1}p^k(1-p)^x$. This gives

$$\frac{\partial}{\partial p}\mu_r = -\mu_{r+1}/q + rk/p^2\mu_{r-1}.\tag{5.33}$$

Cross-multiply and rearrange the expressions to get the result.

Theorem 5.12   *Additivity theorem: If $X_1 \sim \text{NBINO}(n_1, p)$ and $X_2 \sim \text{NBINO}(n_2, p)$ are independent NBINO random variables then $X_1 + X_2 \sim \text{NBINO}(n_1 + n_2, p)$.*

Table 5.1: Properties of negative binomial distribution

Property	Expression	Comments
Range of X	$x = 0, 1,\ldots,\infty$	Discrete, infinite
Mean	$\mu = kq/p$	Need not be integer
Variance	$\sigma^2 = kq/p^2 = \mu/p$	$\mu < \sigma^2$
Mode	$(x - 1), x$	$x = [(q/p)(k - 1)]$ is int.
Skewness	$\gamma_1 = (1 + q)/\sqrt{kq}$	$= (2 - p)/\sqrt{kq}$
Kurtosis	$\beta_2 = 3 + 6/k + p^2/(kq)$	Always leptokurtic
CV	$1/\sqrt{kq}$	
CDF	$F_c(k, p) = I_p(k, c + 1)$	
Mean deviation‡	$2\sum_{x=0}^{\lfloor kq/p \rfloor} I_p(k, x + 1)$	$2k \binom{m+k-1}{m-1} p^{k-1} q^m, m = \lfloor \frac{kq}{p} \rfloor$
Factorial moments	$k^r(q/p)^r = [\Gamma(k + r)/(k)](q/p)^r$	
MGF	$p^k/(1 - qe^t)^k$	$[p/(1 - qe^t)]^k$
PGF	$p^k/(1 - qt)^k$	$p^k/(1 - t + pt)^{-k}$
FMGF	$(1 - qt/p)^{-k}$	
Recurrence	$f(x;k,p)/f(x-1;k,p) = q(k+x-1)/x$	
Tail probability	$\sum_{x>c}\binom{x+k-1}{x} p^k q^x = I_q(c + 1, k)$	$1 - I_p(k, c + 1)$
‡Correction term is $2\delta F(\lfloor kq/p \rfloor)$ where $\delta = kq/p - \lfloor kq/p \rfloor$. Additivity property is $\sum_{i=1}^{m} \text{NBINO}(k_i, p) = \text{NBINO}(\sum_{i=1}^{m} k_i, p)$ for independent random variables.		

Proof. This is most easily proved by the MGF method. We have seen in (5.10) that MGF is $[p/(1 - qe^t)]^k$. As p is the same, replace k by n_1 and n_2 and take the product to get the result. This result can be extended to any number of $\text{NBINO}(r_i, p)$ as follows: $X_i \sim \text{NBINO}(r_i, p)$, then $\sum_i X_i \sim \text{NBINO}(\sum_i r_i, p)$. □

Example 5.13 Data backup   A data transcription company receives 2 GB of data on every working day. New data on their server are backed up daily either to pen drive P1 or P2 with respective probabilities p and $q = 1 - p$ (in other words P1 is selected randomly with probability p for backup) and capacities 100 GB each. When one of the drives become full, let X denote the number of times data were written to the other drive, and Y denote the number of GB (as a multiple of 2) that is free. Find the PMF of X and Y.

Solution: The process of selecting pen drives P1 and P2 can be considered as a Bernoulli process. Suppose P1 is selected in $x + m - 1$ tries, and $(x + m)$th trial also results in P1, so that its capacity is exhausted. Then $X \sim \text{NBINO}(m, p)$. As P1 and P2 are exchangeable (except for the probabilities p and q) the distribution of the number of times P1 was selected when P2 is full is $\text{NBINO}(m, q)$, so that the probability of the number of times either of them is selected when the other one is full is $f(x; k, p) = \binom{x+k-1}{x}(p^k q^x + q^k p^x) = \binom{x+k-1}{x} p^k q^x (1 + (q/p)^k (p/q)^x)$. Obviously x can take the values $0, 1, \ldots, m$. To prove that it is indeed a PMF, consider the relation between $X = \text{NBINO}(k, p)$ and $Y = \text{BINO}(n, p)$ as $\Pr[X \leq r] = \Pr[Y \geq r]$.

If the displaced version of the PMF is used, we get

$$f(y; k, p) = \binom{y-1}{k-1}\left[p^k q^{y+k} + q^k p^{y+k} \right] \quad \text{for} \quad y = k, \; k+1, \ldots. \tag{5.34}$$

This reduces to $\binom{x+k-1}{x}/2^{k+x-1}$ for the first, and $\binom{y-1}{k-1}/2^{y+2k-1}$ for $p = 1/2$.

This problem can be recast in many other forms, some of which are as follows.

1. Suppose two pools of m employees (say contract teachers) are available for hiring. The first pool is supplied by a government agency, and the second pool by a private recruiter. An employer selects from pool 1 with probability p, and from pool 2 with probability $q = 1 - p$ for hiring. If all employees from one pool are exhausted, what is the distribution of the number of employees hired from the other pool?

2. Two teams M and N play a game. Each game is independently won by either M or N (but there is never a tie). Probability of M winning each game is p, and that of N winning is $q = 1 - p$. The game ends as soon as one team wins r (> 0 fixed) games. Let X denote the number of games played by the loser. What is the distribution of X?

3. Two boxes of playing cards C and D contain m cards each. A player randomly picks box C with probability p, and box D with probability $q = 1 - p$. Find the distribution of the number of cards in one box when the other box is empty.

All of these problems are identical in structure. Hence the solution method is the same.

Problem 5.14   A pharma company wishes to conduct human trials for a new medicine. The probability that a randomly selected patient will agree for participation is $p = 0.40$. If 10 patients are needed for human trial, what is the probability that they will be able to get enough patients after asking (i) 12 patients? (ii) 20 patients?

5.6   TAIL PROBABILITIES

As the random variate extends to ∞, the right-tail probabilities are more challenging to evaluate. The left-tail probabilities of $\text{NBINO}(r, p)$ is related to the right-tail probabilities of binomial distribution as $F_k(r, p) = P(X \leq k) = P(Y \geq r) = 1 - \text{BINO}(k + r, p)$. The upper-tail probabilities of a NBINO distribution can be expressed in terms of the incomplete beta function as

$$\sum_{x>c} \binom{x + k - 1}{x} p^k q^x = I_q(c + 1, k). \tag{5.35}$$

The lower-tail probabilities can be found from the complement rule as

$$\sum_{x=0}^{c} \binom{x + k - 1}{x} p^k q^x = I_p(k, c + 1). \tag{5.36}$$

This can also be expressed as tail areas of an F distribution.

Example 5.15   Bounds on SF of negative binomial distribution   Prove that the negative binomial survival function is upper-bounded by $(1 - (x + r)/(x + 1)q) f(x; r, p)$ if $r > 1$, and lower-bounded by the same limit if $r < 1$.

Solution: By definition, SF $S(x) = \sum_{x=k}^{\infty} \binom{x+k-1}{x} p^k q^x$ (we sum from k onwards instead of $k + 1$). Write this in expanded form

$$\binom{x + k - 1}{x} p^k q^x + \binom{x + k - 1}{x + 1} p^k q^{x+1} + \cdots. \tag{5.37}$$

Take $\binom{x+k-1}{x} p^k q^x$ as a common factor and write $\binom{x+k-1}{x+1} = (k - 1)/(x + 1)\binom{x+k-1}{x}$ to get $\binom{x+k-1}{x} p^k q^x [1 + \delta/(x + 1) + \delta^2/(x + 1)(x + 2) + \cdots]$ where $\delta = q(k - 1)$. As $\delta^2/(x + 1)(x + 2) < \delta^2/(x + 1)^2$, and $\delta^3/(x + 1)(x + 2)(x + 3) < \delta^3/(x + 1)^3$, and so on, we get the inequality $S(x) < \binom{x+k-1}{x} p^k q^x [1 + \delta/(x + 1) + \delta^2/(x + 1)^2 + \cdots]$. It is easy to see that the terms in the square brackets is $(1 - \delta/(x + 1))^{-1}$. Thus, $S(x) < \binom{x+k-1}{x} p^k q^x [(1 - \delta/(x + 1))]^{-1}$.

Example 5.16   Negative binomial probabilities   If $X \sim \text{NBINO}(r, p)$, find the following probabilities: (i) X takes even values, and (ii) X takes odd values.

Solution: Let $P_x(t)$ denote the PGF of NBINO(r, p). (i) $P[X$ is even] has PGF given by

$$[P_x(t) + P_x(-t)]/2 = (p^r/2)[1/(1 - qt)^r + 1/(1 + qt)^r]. \tag{5.38}$$

This can be simplified and expanded into an even polynomial in t with the corresponding coefficients giving the desired sum. (ii) The PGF for X taking odd values is $\frac{1}{2}[P_x(t) - P_x(-t)]$. Substitute for $P_x(t)$ to get

$$[P_x(t) + P_x(-t)]/2 = (p^r/2)[1/(1 + qt)^r - 1/(1 - qt)^r]. \tag{5.39}$$

Proceed as above and expand as an odd polynomial in t whose coefficients give desired probabilities.

5.7 TRUNCATED NEGATIVE BINOMIAL DISTRIBUTIONS

Over-dispersed data are often encountered in several applied fields. The Poisson model (next chapter) is not a proper choice for over-dispersed data. Although the zero-truncated and zero-inflated Poisson distributions have been used in such cases, the zero-truncated negative binomial (ZTNB) distribution is an alternative in these cases that gives better fit. These are usually modeled using power-law, Pareto or log-normal distributions. The left-truncated negative binomial distribution is obtained by truncating at a positive integer c.

The resulting PMF is $f(y; k, c, p) =$

$$\binom{y+k-1}{k-1} p^k q^y / \left[1 - p^k \sum_{x=0}^{c} \binom{x+k-1}{k-1} q^x \right] = \binom{y+k-1}{k-1} p^k q^y / I_p(k, c+1),$$

where $I(k, c+1)$ is the incomplete beta function. The alternate form in which Y represents the total number of trials is $f(y; k, c, p) =$

$$\binom{y-1}{k-1} p^k q^{y+k} / \left[1 - p^k \sum_{x=k}^{c} \binom{y-1}{k-1} q^x \right] \quad \text{for} \quad y = k+c, k+c+1, \dots. \tag{5.40}$$

The zero-truncation is a special case in which the PMF becomes

$$f(x; p) = \binom{x+k-1}{k-1} p^k q^x / \left[1 - p^k \right] \quad \text{for} \quad x = 1, 2, 3, \dots. \tag{5.41}$$

An alternate representation is $f(x; n, p) = \binom{x+k-1}{k-1} p^k q^x / (1 - F(0)) = \binom{x+k-1}{k-1} p^k q^x / S(0)$, where $F(x)$ denotes the CDF, and $S(x)$ the survival function, $p^k = F(0)$ (and $1 - F(0) = S(0)$). The PGF can be found easily as

$$P_x(t) = \sum_{x=1}^{\infty} t^x \binom{x+k-1}{k-1} p^k q^x / \left[1 - p^k \right]. \tag{5.42}$$

Take $p^k / (1 - p^k)$ outside the summation, combine t^x and q^x to get

$$P_x(t) = p^k / \left(1 - p^k \right) \sum_{x=1}^{\infty} \binom{x+k-1}{k-1} (qt)^x. \tag{5.43}$$

The sum can be written as $(1 - qt)^{-k} - 1$. This gives the PGF of ZTB distribution as $P_x(t) = p^k[(1 - qt)^{-k} - 1]/(1 - p^k)$. Similarly, the MGF of ZTB follows as $M_x(t) = p^k[(1 - q\exp(t))^{-k} - 1]/(1 - p^k)$.

A right-truncated negative binomial distribution is used in some applications like modeling rank frequencies of graphemes (in linguistics). If truncation is at $n + 1$, the left-out probability is $\sum_{x=n+1}^{\infty} \binom{x+k-1}{k-1} p^k q^x$. This in terms of incomplete beta function is $I_q(n+1, k)$. Thus, the PMF becomes

$$f(x; k, n, p) = [1/(1 - I_q(n+1, k))] \binom{x+k-1}{k-1} p^k q^x$$

$$= [1/I_p(k, n+1)] \binom{x+k-1}{k-1} p^k q^x \tag{5.44}$$

for $x = n + 1, n + 2, \ldots, \infty$. As $\sum_{x=0}^{n} f(x; k, p) + \sum_{x>n} f(x; k, p) = 1$, the above can be alternately expressed as $g(x; k, n, p) = f(x; k, p)/\sum_{x>n} f(x; k, p)$, for $x = n + 1, n + 2, \ldots$.

The moments of truncated negative binomial distribution appear in Gurmu and Trivedi (1992) [30], and a correction to their second moment in Shonkwiler (2016) [98]. If $X_1, X_2, \ldots, X_n$ are IID positive random variables, Shanmugam (1982) [74] obtained a necessary and sufficient condition for these to have ZTNB distributions as $\sum_{j=1}^{k} X_j$ has a joint distribution

$$f(x; k, n, p) = k!/n!r^n s(k, r, n) \prod_{j=1}^{n} \binom{x_j + r - 1}{x_j}, \tag{5.45}$$

for each $k < n$ where $s(k, r, n)$ are Stirling numbers of first kind. A characterization of truncated negative binomial distribution can be found in Shanmugam (1982) [74].

5.7.1   ZERO-INFLATED NEGATIVE BINOMIAL (ZINB) MODELS

The ZINB models are popular for over-dispersed data with excess-zero problem. They extend the classical NBINO models to create two sources of over-dispersion. The PMF can be written as

$$f(x; k, p, \theta) = \begin{cases} 1 - \theta & \text{if } x = 0, \\ \theta\, g(x; k, p) & \text{if } x > 0, \end{cases}$$

where $g(x; k, p)$ is the zero-truncated negative binomial distribution.

5.8   GENERALIZATIONS

Several generalized forms of NBINO(k, p) are available. Consider the PMF

$$f(x; k, n, p) = \frac{n}{n + kx} \binom{n + kx}{x} p^x (1 - p)^{n+kx-x}, \quad x = 0, 1, \ldots. \tag{5.46}$$

for $1 \leq k \leq 1/p, n > 0$. Other popular extensions include the imbalanced inverse binomial (IIB) of Shanmugam (2013) [85] with PMF $f(x; k, p) = \binom{k+x-1}{x} p^x q^k$ where $q = k(1 + \phi)/[(\mu + k)(1 - \phi)]$, $p = 1 - q$; tweaked negative binomial distribution (TNBD) [86], [89], generalized negative binomial distribution of Jain and Consul [35], and Consul and Shenton [22] with PMF $f(x) = (s/t)\binom{t}{x} p^x q^{t-x}$ where $t = s + rx, x = 0, 1, 2, \ldots, 0 < p < -1/r$; and r, s are negative integers.

5.9 APPLICATIONS

Count data are most often modeled using the binomial or Poisson distributions. The mean (np) is always greater than the variance (npq) for a binomial distribution (under-dispersed), whereas the mean and variance are equal for a Poisson distribution (equi-dispersed). But several discrete process exhibit over-dispersion (variance much greater than the mean). Shanmugam and Radhakrishnan (2011) [93] called the ratio $\Pr[X = 1]/\Pr[X = 0]$ as IJR, which for the negative binomial distribution is kq. The IJR value lies in $(0, k)$ as q is a probability. The ratio $V = \sigma_\theta^2/\mu_\theta$ is called excess dispersion index (EDI) where θ denotes the parameter(s) of the distribution. This is related to the EDI as $IJR = k(1 - 1/V)$, or equivalently $V = k/(k - IJR)$. In the above-cited paper, the authors showed (their "n" is our "k", and their over/under dispersion index J is our EDI) that the product V*IJR is the mean, which is a characteristic property of this distribution. Suitable choices in over-dispersion are three parameter Poisson-beta model or negative binomial model. As the negative binomial distribution has an additional parameter that controls the variance ($p\sigma^2 = \mu$), it is a proper choice for over-dispersed data. It is also used in electrochemical processes ([24]), negative binomial regression modeling ([10], [32]), etc.

5.9.1 POWER RESTORATION MODELING

Adverse weather conditions could disrupt electric power distribution systems. This is especially true in tropical regions due to storms and heavy downpour, and in sub-tropical regions due to tornadoes, snowstorms, and blizzards. The failure rate of power distribution systems are modeled either using Poisson or negative binomial models. Power companies are more concerned with power interrupted durations in various places, which may range from a few hours to days. Statistical models built from past data can be used to advantage in mitigating the delay by incorporating the predictors in a negative binomial regression model [32].

5.9.2 CAPTURE-MARK-RECAPTURE MODEL

The capture-mark-recapture (CMR) is a popular technique to estimate an unknown population size. It is used in ecology, marine biology, soil chemistry, entomology, ornithology, and many other fields. Such estimates are used by scientists for conservation and management of endangered species, habitat expansion, predator-prey conflict reduction, predator relocations, etc. A detailed discussion appears in Chapter 7. Assume that the unknown population size is N, among which m are selected at random for marking. The marked items are released and the time interval needed for them to mix well with the population is estimated. After the lapse of time, suppose we continue sampling until k among the m marked items have been sampled. If the event of recapturing a marked item is assumed as a "success," the number of sampling trials needed to get k marked items has a negative binomial distribution $NBINO(k, m/N)$.

5.9.3   MEDICAL SCIENCES

The radiated cells of a cancer patient can be classified as malignant ($x = 1$) or benign ($x = 0$). Among the n possible radiated cells, suppose a random variable X denotes the number of malignant cells given that the patient has carcinogenic metabolism. Then, X conditionally follows a Poisson distribution. Assuming that the non-measurable and latent carcinogenic levels follow a gamma law, the convolution is NBINO distributed (see Example 5.2, page 87). Obviously, zero-radiated cells means that the patient did not undergo any radiation therapy for malignant tumor. Hence a ZINB distribution is more appropriate if the patient group under observation comprises of first-time cancer patients.

5.10   SUMMARY

The negative binomial distribution arises when the probability of a positive result ("success") is sought when there have already been k successes in independent Bernoulli trials. It can also be used to statistically assess the degree to which a discrete distribution is clumped, rather than evenly dispersed.

CHAPTER 6

Poisson Distribution

After finishing the chapter, readers will be able to . . .

- Understand Poisson distribution and its properties.

- Discuss Poisson distribution's relationships with other distributions.

- Comprehend Poisson processes.

- Apply Poisson distribution in practical situations.

6.1 INTRODUCTION

The Poisson distribution was invented by the French mathematician Abraham de Moivre (1718) for $\lambda = 1/2$ [80]. S.D. Poisson (1781–1840) in 1837 obtained the general PMF as the limiting form of a binomial distribution discussed below (page 108). It has a single parameter (usually denoted by Greek letters λ, μ, ν, or θ, and occasionally by uppercase English alphabets (H is sometimes used in genetics, bioinformatics) that denotes the average number of occurrences of an event of interest in a specified time interval (called sampling period in biology, biomedical engineering, and signal processing). It need not be an integer, but must be positive. The λ denotes the counts (arrivals) of random discrete occurrences in a fixed time interval (in temporal processes). It is called *intensity parameter* in some fields like geography, geology, ecology, mining engineering, microscopy; *average vehicle flow* in transportation engineering; *rate parameter* in economics and finance, and as *defect density* in semi-conductor electronics, fiber optics, and several manufacturing systems where events occur along a spatially continuous frame of reference (number of defects in a finished product). However, the occurrence rate for temporal processes is defined as intensity rate = (number of occurrences)/(unit of exposure). This is called the accident rate in highway engineering, mortality rate in medical sciences and vital statistics, and by other names in different fields (e.g., microbe density in clinical pathology).

We denote the distribution by POIS(λ). It can be used to model temporal, spatial (along length or area), or spatio-temporal rare events that are open-ended. Thus, it is also called "law of rarity" or "law of small numbers." For example, it is used to model telephone calls received in a small time interval, accidents (vehicular, airplane, ship, etc.; or accidents or fires in mines, factories, buildings, bridges, and so on) in a fixed time period; automobiles coming at a gas station; insurance claims received in a time period; natural disasters (annual frequency of tropical cyclones and hurricanes, earthquakes), etc. These are all temporal models with different time intervals when the location of occurrence is ignored. They are called *stream models* when a stream of events are simultaneously considered (as in customer check-out counters in supermarkets, vehicular traffic in large cities, check-in counters in railway stations, transmission of data packets in a network, etc. They become spatial processes when different geographical locations (say of the epicenter of natural earthquakes; induced earthquakes due to volcanic eruptions or exceedance of dam capacity (e.g., Kariba dam in Zambia)), or tagging high accident-prone locations in a country map are considered. Shanmugam (2016) ([87], [90]) used the glued bivariate Poisson distribution to come up with a *warning risk index* for earthquakes and aftershocks, that could mitigate future devastations. Other examples of spatial frame of reference include predicting defects in newly manufactured items like clothing sheets, paper rolls or newsprints, cables and wires, micro-chips, etc., where the position of defects are tagged. Applications in agriculture include distribution in space (or time) of plants and ingredients (like moisture or manure), poisonous weeds among regular crops, pest attacks, etc. The quality of an image in image processing applications are determined by spatial resolution, contrast, noise, and artifacts. The overall noise is assumed as Poisson distributed in x-ray, ultrasound, and tomography. As shown below, the number of photons that interact with translucent objects (as it passes through them) is also assumed to be Poisson distributed (in medical imaging, bones absorb more x-rays than muscles so that pixels on the receiver are darker). Similarly, spatial distribution of molecules in an image in molecular spectroscopy and microscopy; disease or crime prevalence in various parts of a city, detection of cracks on the fuselage of an aircraft, etc., are assumed to follow Poisson law. Events are called *point events* in these fields to indicate that they are counts. It is assumed in these fields that the number of event occurrence in closed sub-regions (like cracks in the front portion of an aircraft) depend only on the area, and not on the position of the sub-region. Spatio-temporal applications include predicting earthquakes and tsunamis in a particular region over a time period, outbreak of epidemics in a geographical region over a time period, surface modeling techniques in electrical engineering that involve positive and negative charge separation resulting in electrostatic potential/field forming on the interface, and heterocoagulation techniques used in water treatment and hydrology. Ecology applications include predator-prey conflicts, species extinction, deforestation of areas, occurrences of wild fires, melting of polar ice-caps, and so on in specified locations over a time period. Traffic engineering uses the Poisson model when the vehicular traffic flow is low. As traffic signals, highway humps, pedestrian crossings, etc., can cause traffic flow disturbances, they may be inappropriate in the vicinity of such objects for stream modeling. Similarly, deterministic flow in which vehicle arrivals or departures are uniform (as in trains, airlines, ships with fixed time-tables; or rush-hour traffic on highways during which vehicles maintain more or less the same speed) cannot be modeled by a Poisson process. But point processes may apply provided that the flow is non-uniform. This means that we could still develop a Poisson model for the number of vehicles arriving randomly at a isolated traffic signals (away from other signals and interruptions like speed breakers), or passing through a multi-lane hump in a small time interval.

It has also been applied to volumetric data (both static and dynamic). For example, the density (expressed as counts) of micro-organisms in water or other liquids in small-volume samples, harmful bacteria in fruits, vegetables, and other food for microbial risk assessment, geo-referenced rare count data, etc., can be modeled either with Poisson (when data are equidispersed), negative binomial, or three parameter Poisson-beta distribution (data are overdispersed) distributions. Thus, the number of harmful microbes present in a unit volume of water is assumed as Poisson distributed. Poisson distribution may also be applied to flows of liquids (in motion). They are characterized by nonuniformity of the substance to be modeled in a dynamically collected sample. Not only rare events, but some recurrent events also are modeled using Poisson distribution. Consider a country with lots of villages. The power load suddenly goes up around 6 PM in the evening, and tails off to a low level after 10 PM or later when demand drops in most of the villages (sleep mode). Then, it may again peak during early morning hours (say 6 AM), but the tailing off is much slower (except in countries where alternate power sources like solar power, wind power, etc., are in use) since power is consumed during the entire day in offices, factories, hospitals, and educational institutions in addition to domestic demand. Thus, an approximate Poisson model may be appropriate in rural areas, on a recurring basis.

6.1.1 IMPORTANCE OF COUNTS

Poisson distribution arises from count data. How the counts are obtained depends on the field of application. This may not be so important for a statistician, but modelers and researchers in different fields should know what is the best method to obtain the counts. A statistician is more concerned with the types and accuracy of the count data, rather than the method used to obtain it. The counts may be obtained either visually, using data collection devices, using signal or image processing techniques, using properties of materials, using special sensors or a combination of these. For example, inspecting a seismograph and counting the peaks above a cutoff can be used to find the number of earthquakes of specific intensities (say above 6 on Richter scale). Devices that obtain data directly from a location where an event or transaction occur (barcode readers, fastags, mobile devices or Apps, ADC counts recorded by a detection system) are sometimes used to obtain the counts. Number of deaths occurring due to rare diseases can be obtained from hospitals or government databases. Very small sensors (called probes) in scanning-tunneling electron microscopy are used to count electrons passing through a gap between two conducting surfaces. The Coulter counter used in hematology uses the magnitude of voltage differences between two electrodes when non-conducting cells (like RBC) pass through a small aperture (typically $100\ \mu$m), and the change in resistance is converted into counts. Light-scattering properties (fluorescent probes) are used in crystallography and spectroscopy, whereas laser beams are used in semiconductor industry to obtain counts. In some applications like food-processing and pharmaceuticals, an equal amount of another liquid without microbes may be mixed in to double the volume. In this case the Poisson mean gets scaled by half. A problem encountered is that the contamination can be either random (well mixed, unstructured, or well cooked) or localized (in one or more clusters).[1] A grid (called electrophoresis column in bioinstrumentation and spectroscopy) is the unit used in spatial processes for obtaining the counts. Grids are closed and bounded geometric regions, which are usually rectangular, square, or hexagonal in shape (for 2D) and more or less of equal area (or volume). If the localized region is much smaller (as in nanotechnology), a smaller grid size may have to be used for Poisson modeling. As the Poisson distribution approaches a bell-shaped curve for large λ values, it is more appropriate to model rare events when the average is small. This is the reason behind the dilution technique mentioned above (this is further discussed in page 122). Another solution is to use a mixture of distributions that may include discrete and continuous (e.g., log-normal) distributions. This is usually the case in clinical pathology and bacteriology when one organism dominates over the others and we wish to build a single model for the entire population.

It is also used in many engineering fields as shown below. The unit of the time period (or space for spatial data) in these cases are implicitly assumed by the modeler, and could depend upon other variables. For instance, cracks and leakages of aircraft engines are rare events for which the time-interval depends on the age (total hours flown) of the aircraft (or last replacement date of the engine or parts thereof). Similarly, some of the medical conditions observed in a country could be large, so that when restricted to smaller regions, towns, or hospitals will result in a reasonably rare event for which an appropriate time interval can be chosen, and multiple Poisson models developed for various non-overlapping larger regions. A wrong choice of the time period or space may lead to convoluted Poisson models. But always keep in mind that in temporal processes, the λ denotes event occurrences per interval multiplied by the interval length (if the interval length is unity, it represents the number of occurrences of events in an interval), number of events per grid multiplied by grid size in the case of 2D spatial processes (also called $\mathbf{R}^2$ spatial process). The mean in this case is assumed to be $c\lambda$ where λ is the expected number of points per unit area, and c is the area of a closed uniform region of interest (usually a rectangular grid).

When the frame of reference is 2D or higher dimensional space, this represents the number of occurrences of the event in a properly aligned *grid*. For example, radioactive decay occurs at a constant rate when temperature and pressure are kept constant. Alpha particle (helium nucleus with two each of protons and neutrons) emissions in short time intervals, photoelectric effect, and electron multiplication in a PMT tube can all be modeled using a Poisson law. Usually, α particles emitted by a radioactive element is captured using scintillation counters that count the oscillations, or exposed on a circular plate (coated with some chemicals like zinc sulphide) which is divided into sub-regions of unit size in such a way that two or more particles do not strike a

---

[1]Stratified sampling is usually used when distribution of microbes are localized.

unit region simultaneously (alternatively, the receiver can be placed at such a distance that the entire unit receives at most one particle at a time, and the time unit is adjusted accordingly, with the distance decreased at a constant rate as the decayed particle reduces in size).[2] The striking α particles produce a speck of light on the screen which can be counted by sensors. The counts can then be modeled as $\exp(-\text{rate} \times \text{time})$ $(\text{rate} \times \text{time})^x/x!$, for $x = 0, 1, 2, \ldots$. Here, the rate denotes the count which depends on the particle (say 1 per 2 seconds for Polonium).

The above discussion shows that some *technical knowhow* in the modeling field, and accurate information regarding the time interval, space area, or volume of samples is absolutely necessary as it could definitely improve the modeling process. One example is in the chip-defects in semi-conductor industry. Unless it is known whether the counts represent random or pattern errors (or both together), the modeler may come up with a spurious Poisson distribution. It may also be noted that a Poisson model may not provide conclusive evidence about the spatial properties of a process (like defect patterns on wafers) as the counts within a grid are used without regard to their point position coordinates.

6.2   PROBABILITY MASS FUNCTION

The PMF of a Poisson distribution POIS(λ) is given by

$$p_x(\lambda) = e^{-\lambda}\lambda^x/x!, \quad x = 0, 1, 2, \ldots, \tag{6.1}$$

where $e = 2.71828$ is the Euler's constant (base of the natural logarithm), and λ is the average number of events that occur in a fixed time interval (see section on Poisson process in page 122). Some authors use n in place of x in the PMF to indicate that the values taken are counts, while others write the PMF as $p_x(\lambda) = [1/e^\lambda][\lambda^x/x!]$. We will stick with the x notation because it is used with other discrete distributions like binomial, negative binomial and geometric distributions. Obviously, summing over the range of x values gives $\sum_{x=0}^{\infty} e^{-\lambda}\lambda^x/x! = e^{-\lambda}\left(1 + \lambda^2/2! + \cdots\right) = e^{-\lambda}e^\lambda = 1$. By writing λ^x as $\exp(x\log(\lambda))$, and $1/x!$ as $h(x)$ it is easy to show that it belongs to the exponential family. The $\lambda = 1$ is a special case in which $f(x) = 1/(e\,x!)$.

In very large scale integrated-circuit (VLSI) design, the yield of a chip is expressed as the probability of no defects.[3] If defects are assumed to follow the Poisson distribution, the yield is given by $P(x = 0) = \exp(-\lambda)$ under the assumption that defects occur independently, and

---

[2]It can also be tunneled through a conic tube (e.g., GM tube) and a receiver kept in the field of view.
[3]Percentage of wafers that reach the final probe is called wafer yield.

the entire die area is taken as the unit. It is known as first time yield (FTY) in manufacturing engineering. This is called the Poisson yield model, and is discussed later.

Example 6.1   Radioactive emissions   If the time intervals during which two alpha particles are observed is twice the time intervals in which only one alpha particle is observed among 1640 intervals, find the number of intervals in which no alpha particle is emitted.

Solution: Assume that alpha particle emissions is approximately Poisson distributed. Let X denote the number of alpha particles emitted in a fixed time interval. It is given that $\Pr[x = 2] = 2 * \Pr[x = 1]$ (because probabilities are taken as relative frequencies). As $X \sim$ POIS(λ), we have $\exp(-\lambda\ \lambda^2/2!) = 2\exp(-\lambda\ \lambda/1!)$. Cancel out common factors to get $\lambda/2 = 2$ or $\lambda = 4$. From this the number of intervals in which no alpha particles are emitted is $\exp(-\lambda)\lambda^0/0! = \exp(-4) = 0.0183156$. The number of intervals in which no alpha particle is emitted is $0.0183156. * 1640 = \lfloor 30.0376 \rfloor = 30$.

Example 6.2   Poisson PMF   A researcher tells you that 540 vehicles per hour pass through a two-way intersection during peak hours with 300 vehicles in east-west lane, and the rest in north-south lane. You have 20 seconds to verify the claim. What is the probability that you will see (i) no vehicles in both lanes, (ii) two or more vehicles in east-west lane? (iii) exactly one vehicle in north-south lane?

Solution: As the lanes are not distinguished in part (i) we take $\lambda = 540/(3 * 60) = 3$ in a 20 s time period. Probability that no vehicles pass through during this time is $p(0) = \exp(-3)3^0/0! = 0.049787$. This is called vehicle headway (h) in transportation engineering, which for single-lane highways and roads is given by $\Pr[h \geq t] = p(0)$ (headway is the time interval between two successive vehicles, which is usually measured by the front bump part due to a variety of vehicles that passes through a highway). In part (ii), we are given that 300 vehicles pass through in one hour so that on the average $300/60 = 5$ vehicles are to be expected per minute. Thus, $\lambda = 5/3$ for a 20 s interval that we observe. This implies that $p(0) + p(1) = \exp(-5/3)[1 + 5/3] = 0.503668$. Hence, the probability that you will see two or more vehicles in east-west lane is $1 - [p(0) + p(1)] = 0.496332$. As there are 240 vehicles passing in north-south lane, $\lambda = 240/(60 * 3) = 4/3$. Hence, probability of finding exactly one vehicle is $\exp(-\lambda)\lambda^1/1! = \exp(-4/3) * 4/3 = 0.351463$.

In some applications in engineering, astronomy, geology, clinical pathology, etc., the time interval t or space s (of observed area, volume, etc.) are kept as generic [1]. This results in the *rate and time* model

$$f(x; \lambda) = \exp(-\lambda t)(\lambda t)^x/x! \tag{6.2}$$

for temporal events. This is just the model given above except that the average number of occurrences has been dilated to a new interval of time (or space) t' as $\mu = \lambda t$.

6.3   DILUTION TECHNIQUES

Dilution is similar in principle to the parameter dilation discussed previously (which is essentially one dimensional as time is the variable). Clinical pathology, microbiology, and many engineering fields that use properties of fluids, liquids, and gases use "dilution techniques" to estimate counts to a desired accuracy. Replace t in the above model by s for spatial (2D) or v for volumetric (3D) processes. Bacteriologists typically dilute the sample by a factor of 10^{-4} to 10^{-5} (called dilution factor), so as to rarify it (forcibly make it rare). It is stirred to make it uniform. Some microbiological tests take samples (say of 1 mm^3) and spread it uniformly onto microscopic plates containing nutrients for the sample to grow and reproduce. Multiple replicates may be taken in sensitive tests to account for sampling variability. Colonies of bacteria will appear on the plates after some time. A proper grid size is then used to count the number of colonies in each grid. The mean colony count can be used to estimate bacterial density in the original (undiluted) medium by multiplying by the dilution factor. Clinical pathologists also use such techniques (dilution and condensing) when the expected count is too large (e.g., urinalysis for protein molecules detection) or too low. If unit volume of liquid containing λ microbes is diluted with m units of another liquid without microbes, the resulting mixture will be Poisson distributed with mean $\lambda/(m+1)$. Similarly, if m units of a liquid with λ microbes on the average is mixed with n units of another liquid with θ microbes, the resulting liquid is assumed to have $(m\lambda + n\theta)/(m+n)$ microbes on the average, provided that the mixing is uniform.

6.4   MOMENT APPROXIMATION

The binomial distribution BINO(n, p) for $n > 2$ can be approximated by the Poisson law, where n is a positive integer (number of trials) and $0 < p < 1$ is the probability of success in each trial. As a first try, consider the mean and variance of binomial and Poisson distributions. It is shown below that both the mean and variance are λ for the Poisson distribution. Poisson distribution assumes that event occurrences are rare. This means that the probability of an event occurring at any instant in time $[t, t+dt)$ is negligibly small. It does not tell us anything about the magnitude of λ. This has the practical implication that when the data arising from *rare events* have the same (or almost the same) mean and variance, then a Poisson model to fit the data is more likely to be a good choice. This observation was used by Student (1907) among other researchers. The mean of binomial distribution is $\mu = np$, and the variance is $\sigma^2 = np(1-p) = \mu q$ where n is the number of trials, p is the fixed probability of success, and $q = 1 - p$. If the Poisson distribution is to be a good match (approximant) of binomial distribution, the mean and variance of them should be very close to each other. This means that np $\simeq np(1-p)$ (because the mean and variance are equal for the Poisson distribution) where $\simeq$ denotes *approximately equal*. Cancel out np on both sides to get $1 \simeq 1 - p$. This implies that p is VERY SMALL. This does not mean that the binomial distribution always tends to the Poisson law when p is small because we have equalized only the means (center of gravity) and variance (spread). Equating the means and variances gives only a crude approximation for two distributions. This fact was used by many researchers, including Patnaik(1949) [62] to approximating a noncentral χ^2 distribution using a central χ^2 distribution. Many researchers have applied this technique to various statistical distributions subsequently. In the case of dependent events (where the probability of success is p_i), Arratia, Goldstein, and Gordon (1989) [3] states that "convergence to the Poisson distribution can often be established by equating only the first and second moments, but not higher-order ones." To improve upon the approximation, there are still other parameters to be considered (like skewness and kurtosis and higher order moments). This is because the binomial distribution is skewed only when p is very small (close to zero) or very large (close to 1), and not otherwise. Thus, a *necessary condition* for a binomial distribution to tend to the Poisson law is that p (or $q = 1 - p$) must be small.

6.5 DERIVATION OF POISSON LAW

Poisson (1837) derived the Poisson distribution as a limiting case of binomial distribution. Symbolically,

$$\lim_{n \to \infty} \binom{n}{x} p^x (1 - p)^{n-x} = e^{-\lambda} \lambda^x / x!, \tag{6.3}$$

where $\lambda = np$ is the parameter of Poisson law. Consider the binomial distribution BINO(n, p) with PMF

$$f_x(n, p) = \binom{n}{x} p^x (1 - p)^{n-x}, \tag{6.4}$$

where n is a positive integer. The PGF is found in Chapter 2 as

$$P_x(n, p; t) = (q + pt)^n, \tag{6.5}$$

where $q = 1 - p$. Take log of both sides to get

$$\log (P_x(n, p; t)) = n \log(q + pt). \tag{6.6}$$

Write q as $1 - p$ and n as $-(-n)$ to get

$$\log(P_x(n, p; t)) = -n(-\log(1 - p(1 - t))). \tag{6.7}$$

Expand using

$$-\log(1 - x) = x + x^2/2 + x^3/3 + \cdots \tag{6.8}$$

to get

$$\log (P_x(n, p; t)) = -n \left[p(1 - t) + p^2(1 - t)^2/2 + p^3(1 - t)^3/3 + \cdots \right]. \tag{6.9}$$

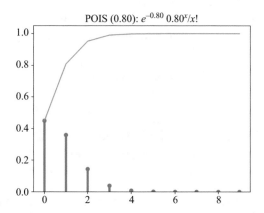

Figure 6.1: PMF and CDF of POIS(0.80).

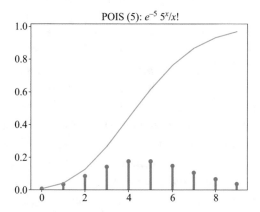

Figure 6.2: PMF and CDF of POIS(5.00).

Note that the binomial distribution has two independent parameters n and p. There are different limiting cases to be considered (i) $n \to \infty$ faster than $p \to 0$, (ii) $p \to 0$ faster than $n \to \infty$, and (iii) $n \to \infty$ and $p \to 0$ in a controlled way (say such that np remains either in a fixed interval $[a < np < b]$ or remains a constant λ). Write $np = \lambda$ in the first term and take the negative sign inside the bracket. Then RHS reduces to $\lambda(t - 1) - np^2(1 - t)^2/2 + \cdots$. When the RHS is exponentiated, the first term becomes $\exp(\lambda(t - 1))$ which is the PGF of Poisson distribution. Collect the constant coefficients from each of the terms in (6.9) to get

$$C = -\left[np^2/2 + np^3/3 + np^4/4 + \cdots\right]. \tag{6.10}$$

Discarding terms other than the first shows that the convergence is dependent on np^2. This gives a better expression that throws more light on the convergence:

$$\binom{n}{x} p^x (1-p)^{n-x} = \exp(-np)\,(np)^x/x! + O\left(np^2\right), \tag{6.11}$$

where $O()$ denotes the Big-O notation [48]. The rest of the terms contain higher-order powers of p. Hence, in the limit the binomial distribution tends to the Poisson law. The implied meaning of this equation is that the convergence is dependent linearly on n and quadratically on p. In other words,

``The binomial distribution tends to the Poisson law when p tends to zero faster than n tends to infinity.''

This allows us to approximate the binomial probabilities by the Poisson probabilities even for very small n values (say $n < 10$), provided that p is comparatively small. Of course, the above result holds if p is taken as λ/n, which falls under case (iii). However, the error in approximation is small when λ is small. An implied meaning of this is that λ is naturally small when p tends to zero faster than n tends to infinity. As n and p are always positive, their product $\lambda = np$ is always ≥ 0. A correction must be applied in the other cases discussed above. For instance $\lambda = np$ can be large when n is very large and p is relatively not so small. If $\lambda < 1$, a constant multiplier $e^{\lambda p/2}$ applied to the computed Poisson approximation can improve upon the binomial probability towards the middle (x values near np). Thus $b(x; n,p) \cong P(\lambda) \times e^{\lambda p/2}$ where $\lambda = n * p < 1$.

As the binomial distribution involves binomial coefficients $\binom{n}{x}$ that satisfies the symmetry relationship $\binom{n}{x} = \binom{n}{n-x}$ we could approximate the binomial distribution when p tends to 1 (certainty), as $\text{BINO}(n, q)$ where $q = 1 - p$ is small using a Poisson law with $\lambda = nq$. This means that the Poisson law is applicable not only for rare events but for certain events as well. This can be understood in a better way by considering the relationship between Poisson and Bernoulli distributions. If multiple events occur in time t (for temporal) or space s (for spatial and higher dimensional) of a Poisson process, we must sub-divide the interval into finer granularity such that only one event may occur in the smaller interval or region so that it results in a Bernoulli process (which can be modeled by a Bernoulli distribution as $f(x, p) = 1$ if an event occurs in the sub-interval and 0 otherwise).

Consider Equation (2.67) in page 44, which is reproduced below:

$$\log\left(P_x(t; n, p)\right) = -n\left[p(1-t) + p^2(1-t)^2/2 + p^3(1-t)^3/3 + \cdots\right]. \tag{6.12}$$

Keeping the first term intact, and collecting constant terms from the rest, we get the RHS as

$$np(t-1) - \left[np^2/2 + np^3/3 + \cdots\right] = np(t-1) - n\left[-\log(1-p) - p\right]$$
$$= np(t-1) + n[\log q + p] = \lambda(t-1) + \lambda + n\log(q). \tag{6.13}$$

Exponentiating LHS and RHS, we see that the first term becomes the PGF of the Poisson distribution.

6.6   PROPERTIES OF POISSON DISTRIBUTION

This distribution has a single parameter λ, which is both the mean and variance. The PMF is easy to compute for small λ values. It is an excellent choice for forming mixture distributions (like noncentral χ^2 distribution). The difference of two independent Poisson random variables has the Skellam distribution with PMF

$$f(x, \lambda_1, \lambda_2) = e^{-(\lambda_1 + \lambda_2)} (\lambda_1/\lambda_2)^{x/2} I_x \left(2\sqrt{\lambda_1 \lambda_2}\right), \tag{6.14}$$

where $I_x()$ is the modified Bessel function of the first kind.

Problem 6.3   A call center operator can handle a maximum of four calls per minute. If the number of calls coming is Poisson distributed with mean 3, what is the probability that (i) operator is idle for the next 2 min, and (ii) there are calls waiting in the queue for next one minute?

6.6.1   MOMENTS AND MGF

The first moment is readily obtained as

$$\mu = E(X) = \sum_{x=0}^{\infty} x e^{-\lambda} \lambda^x / x! = \lambda e^{-\lambda} \sum_{x=1}^{\infty} \lambda^{x-1}/(x-1)! = \lambda e^{-\lambda} e^{+\lambda}. \tag{6.15}$$

Using $e^m * e^n = e^{m+n}$ this reduces to λ. The quantity $1/\lambda$ is known by different names like "mean return period" in natural disaster fields like earthquake engineering, epidemiology, etc. To find the second moment $E(X^2)$, write x^2 as $x * (x-1) + x$ to get

$$E\left(X^2\right) = \sum_{x=0}^{\infty} x^2 * e^{-\lambda} \lambda^x / x! = \sum_{x=0}^{\infty} [x(x-1) + x] * e^{-\lambda} \lambda^x / x! = \lambda^2 + \lambda. \tag{6.16}$$

From this the variance is found as $V(X) = E(X^2) - E(X)^2 = \lambda$ (see Table 6.4). An implication of this is that the variance can be estimated from the overall mean of a random sample. Although the count steadily decreases for increasing value of x, there are also left-skewed Poisson distribution in which the frequency of occurrence is in reverse. Consider the mortality rate of coronavirus (COVID-19) infected people. Data available at the time of this writing appears in the Table 6.1. As is easy to notice, the death rate is high for octogenarians (and data are unavailable for older people). A Poisson distribution can be fitted by assuming $x = 0$ for people older than 80, $x = 1$ for 70–79 age group, and so on, and taking the percentage as counts. As the data pertains to dead patients in various age groups, the total count is $N = 18 + 9.8 + 4.6 + 1.3 + .4 + .18 + .09 + .02 = 34.39$. This gives the mean as

Table 6.1: Mortality rate of coronavirus (COVID-19)

Age Group	Mortality Rate (%)	Expected Value (%)
≥80	18	16.109
70–<80	9.8	12.217
60–<70	4.6	4.632
50–<60	1.3	1.171
40–<50	0.40	0.222
30–<40	0.18	0.034
20–<30	0.09	0.004
10–<20	0.02	0.00046
<10	0.00	.000048
Total	34.39	34.38995

$(9.8 + 2 * 4.6 + 3 * 1.3 + 4 * .4 + 5 * .18 + 6 * .09 + 7 * 0.02)/N = 0.758359988$. The fitted values are shown in column 3. The discrepancy can partially be attributed to the fact that we have used percentages instead of counts to fit the model. An alternate choice is the Marshall–Olkin discrete uniform distribution.

6.6.2   FACTORIAL MOMENTS

Factorial moments of a Poisson distribution are easier to find because of the presence of $x!$ in the denominator of the PMF. The rth factorial moment is

$$\mu_{(r)} = E\left[x_{(r)}\right] = E[x(x-1)\dots(x-r+1)] = \sum_{x=0}^{\infty} x(x-1)\dots(x-r+1)e^{-\lambda}\lambda^x/x!$$

$$= \lambda^r e^{-\lambda} \sum_{x=r}^{\infty} \lambda^{x-r}/(x-r)! = \lambda^r e^{-\lambda}e^{+\lambda} = \lambda^r. \tag{6.17}$$

Higher-order moments can be obtained from this as $\mu_2' = \lambda^2 + \lambda$, $\mu_3' = \lambda^3 + 3\lambda^2 + \lambda$.

As normality is approached when the mean becomes large, the CDF can be approximated as

$$F(x;\lambda) = \Phi\left((x + 0.5 - \lambda)/\sqrt{\lambda}\right), \tag{6.18}$$

where $\Phi()$ is the standard normal CDF. The first inverse moment is used in econometrics. As the support is from 0 to ∞, we define it as $E(1/(X+1))$ to avoid the possibility of divide by zero. This is given by $\sum_{x=0}^{\infty}(1/(x+1))\exp(-\lambda)\lambda^x/x!$. Multiply and divide by λ and take $\exp(-\lambda)$ outside the summation to get $\exp(-\lambda)/\lambda \sum_{x=0}^{\infty} \lambda^{x+1}/(x+1)! = \exp(-\lambda)/\lambda(\exp(\lambda) - 1)$. This simplifies to $(1 - \exp(-\lambda))/\lambda$. This can be approximated as $1/\lambda$ for large λ values.

Table 6.2: Properties of Poisson distribution

Property	Expression	Comments
Range of X	$x = 0, 1,\ldots,\infty$	Discrete, infinite
Mean	$\mu = \lambda$	Real number
Variance	$\sigma^2 = \lambda$	$\Rightarrow \mu = \sigma^2$
Mode	$[\lambda - 1, \lambda]$, if λ is integer	$\lfloor \lambda \rfloor$ if not integer
Skewness	$\gamma_1 = 1/\sqrt{\lambda}$	
Kurtosis	$\beta_2 = 3 + 1/\lambda$	Leptokurtic always
Even sum	$\frac{1}{2}(1 + e^{-2\lambda})$	
CDF	$P[x \leq r] = \frac{1}{\Gamma(r+1)} \int_\lambda^\infty e^{-x} x^r dx$	
CV	$\sqrt{\lambda}$	
Mean deviation	$2\sum_{x=0}^{\lfloor \lambda \rfloor - 1} \gamma(x + 1, \lambda)/\Gamma(x + 1)$	$\frac{2}{\lfloor -\lambda \rfloor!} \exp(-\lambda)\lambda^{\lfloor \lambda \rfloor + 1}$
Moments	$\mu_r = \lambda\sum_{i=0}^{\lfloor r-2 \rfloor}\binom{r-1}{i}\mu_i$	$r>1$, $\mu_0 = 1$
r^{th} cumulant	λ	
Factorial moments	λ^r	
FMGF = $E(1 + t)x$	$\exp(t\lambda)$	
MGF	$e^{\lambda(e^t-1)}$	
PGF	$e^{\lambda(t-1)}$	
Additivity	$\sum_{i=1}^m \text{POIS}(\lambda_i) = \text{POIS}(\sum_{i=1}^m \lambda_i)$	Independent
Recurrence	$f(x;n,p)/f(x-1;n,p) = \lambda/x$	
Tail probability	$\sum_{x=0}^m (\frac{e^{-\lambda}\lambda^x}{x!}) = \frac{\gamma(m+1,\lambda)}{\Gamma(m+1)}$	$\gamma(m, \lambda) = \int_\lambda^\infty e^{-y}y^{m-1}dy$

Approaches normality when $\lambda \to \infty$. $\gamma(m, \lambda)$ is the incomplete gamma function.

6.6.3   MOMENT GENERATING FUNCTION

The MGF is

$$M_x(t) = E\left[e^{tx}\right] = \sum_{x=0}^\infty e^{tx} e^{-\lambda}\lambda^x/x! = e^{-\lambda}\sum_{x=0}^\infty \left(\lambda e^t\right)^x/x! = e^{-\lambda}e^{\lambda e^t} = e^{\lambda(e^t-1)}.$$

From this the PGF is obtained by replacing e^t by t as $P_x(t) = e^{\lambda(t-1)}$. See Chattamvelli and Shanmugam (2019) for other GFs.

Example 6.4 Mode of Poisson distribution   Prove that the mode of the Poisson distribution is $\lfloor \lambda \rfloor$ if λ is non-integer, and is bi-modal with the modes located at $[\lambda - 1, \ \lambda]$ otherwise.

Table 6.3: Mean deviation of Poisson distribution using our power method for λ integer

λ	Direct	1	2	3	4	5	6	7	8	9
5	1.755	.0135	.0943	.3436	.8737	1.755				
7	2.086	.0018	.0164	.0757	.2392	.5852	1.186	2.086		
8	2.233	.0007	.0067	.0342	.1190	.3182	.7007	1.328	2.233	
10	2.502	.0012	.0066	.0273	.0858	.220	.480	.921	1.586	2.502

First column gives λ values of Poisson distribution. Second column onward are the values accumulated using (6.30) on page 118. Row for $\lambda = 10$ has been left shifted by one column.

Table 6.4: Variance of discrete distributions

Distribution	Mean	Variance	σ^2 as μ	Ordering
Binomial	np	npq	μq	$\sigma^2 < \mu$
Poisson	λ	λ	μ	$\sigma^2 = \mu$
Geometric	q/p	q/p^2	μ/p	$\sigma^2 > \mu$
Negative Binomial	kq/p	kq/p^2	μ/p	$\sigma^2 > \mu$
Hypergeometric	$\frac{nk}{N} = t$	$\frac{t}{N-1}(1-k/N)(N-n)$	$\frac{\mu}{N-1}(1-k/N)(N-n)$	$\sigma^2 < \mu$
Neg. Hypergeometric	$r\frac{N+1}{k+1} = t$	$\frac{t(N-k)(k+1-r)}{(k+1)(k+2)}$	$\mu\frac{N-k}{(k+2)}(1-\frac{r}{(k+1)})$	$\sigma^2 > \mu$
Uniform	$(N+1)/2$	$(N^2-1)/12$	$\mu(\mu-1)/3$	
Logarithmic	$q/[-p \ln(p)]$	$\mu(1/p-\mu)$	$\mu(\mu-1)/3$	$\sigma^2 > \mu$

For discrete uniform, hypergeometric, and negative hypergeometric distributions, the inequality depends on parameter values. For discrete uniform distribution $\sigma^2 < \mu$ when $N < 7$ and $\sigma^2 > \mu$ for $N > 7$. They are equal when $N = 7$.

Solution: Consider the ratio $f_x(k, \lambda)/f_x(k-1, \lambda) = \lambda/k$. If $k \leq \lambda$, the LHS is strictly increasing. Otherwise it is strictly decreasing. If λ is integer, λ/k will assume the last integer value at $k = \lambda$ (if λ is a prime number, this occurs only once for $k > 1$, but if it is composite, the ratio could be integer for more than one value of k). Thus, if λ is an integer, the RHS becomes 1 when $k = \lambda$ giving $f_x(\lambda, \lambda) = f_x(\lambda - 1, \lambda)$ (we have simply substituted $k = \lambda$). Thus, the maximum occurs at $k = \lambda - 1$ and λ. Otherwise there is a single mode at $[\lambda]$, the integer part (Table 6.2).

Coefficient of variation (CV) is one of the measures of randomness of a distribution. The CV of Poisson distribution is $\sigma/\mu = \sqrt{\lambda}/\lambda = 1/\sqrt{\lambda}$, which is a decreasing function in λ for $\lambda > 1$.

Example 6.5   Defectives in shipment   Consider a collection of items like light bulbs, transistors, of which some are known to be defective with probability $p = .001$. Let the number of defectives in a shipment follow a Poisson law with parameter λ. How is p and λ related? What is the probability of finding (i) no defectives, or (ii) at least 2 defective items in a shipment containing 20 items?

Solution: If n is the number of items in the shipment, p and λ are related as $np = \lambda$. To find the probability of at least two defectives, we use the complement-and-conquer rule. The complement event is that of finding either 0 or 1 defective. The corresponding probabilities are $e^{-\lambda}$ and $\lambda e^{-\lambda}$. As $n = 20$, $n * p = 20 * .001 = 0.02$; (i) probability of finding no defectives $= e^{-0.02} = 0.98019867$, and (ii) substitute for λ to get $e^{-0.02} + 0.02 * e^{-0.02} = 0.9801986 + 0.0196039 = 0.9998$ as the complement probability. From this the required answer follows as $1 - 0.9998 = 0.0002$.

6.6.4   ADDITIVITY PROPERTY

If $X_1 \sim \text{POIS}(\lambda_1)$ and $X_2 \sim \text{POIS}(\lambda_2)$ are independent, then $X_1 + X_2 \sim \text{POIS}(\lambda_1 + \lambda_2)$. This is most easily proved using the MGF. Using $M_{X_1+X_2}(t) = M_{X_1}(t) * M_{X_2}(t)$, we get $M_{X_1+X_2}(t) = e^{(\lambda_1+\lambda_2)(e^t-1)}$. This result can be extended to an arbitrary number of random variables. An immediate consequence of this property is that if the sum of n independent random variables is Poisson distributed, then each of them is also Poisson distributed. In particular, if $X_k \sim \text{IID POIS}(\lambda/n)$ for $k = 1, 2, \ldots, n$; then $S = X_1 + X_2 + \cdots + X_n \sim \text{POIS}(\lambda)$. Such models are used in actuarial sciences as follows: Suppose a company offers insurance portfolio, and N_k denotes the number of claims received in year k which has a Poisson distribution $\text{POIS}(\lambda)$. The total claims for n years is $N_1 + N_2 + \cdots + N_n = M$ (say). If year-to-year claims are assumed to be independent, then $M \sim \text{POIS}(n\lambda)$. This shows that $M/n \sim \text{POIS}(\lambda)$ so that for large n we could apply the central limit theorem to approximate it by a normal distribution.

Splitting an already-known Poisson process may sometimes give more insight. For example, signal plus noise processes are sometimes assumed to be the sum of two Poisson random variables. Consider the number of traffic accidents at a specific intersection. If it is known to be Poisson distributed, we could further break it into single vehicle accident, two-vehicles accident, and multi-vehicles accident.[4] This property is used in several engineering applications like radioactivity detection on planet surfaces or deep water, microwave source detection in deep-sky

---

[4]Most of the accidents at four-way intersections involve two vehicles, so that single-vehicle and multi-vehicles accidents are rare. Similarly, accidents involving vehicles and pedestrians, pets, cyclists, etc., depend on whether the intersection is at the center of a city or away from it.

scans, etc. In such cases, there could always exist a background noise which might be significant. For example, a small amount of radioactivity is always present on Earth's surface due to cosmic background radiation penetrating the atmosphere. Similarly, *thermal noise* is always present in radar receiving antenna that represents electromagnetic (EM) waves produced by random thermal motion of charged particles in the atmosphere and in the vicinity, in addition to its own thermal noise due to randomly moving.[5] If the background radiation λ_b is accurately estimated by averaging multiple geographically spread sensor data at different time intervals, a parametric decomposition of λ as $\lambda = \lambda_b + \lambda'$ where λ' is the actual radioactive source allows us to model it, and identify the source precisely. The value of λ' is zero when there is no radioactive source present in the vicinity of the detector, and it will peak when we move toward a single radioactive source. Theoretically, this can be modeled as a convolution of two Poisson laws. Multiple sources of radiation present in close proximity (as in unexploded or discarded bombs or machinery that works on nuclear energy) will give rise to multiple convolutions. Poisson kriging is an interpolation method used to predict radiation levels at unknown spatial locations using known levels at geo-tagged (usually multiple) locations under the assumption that data follows a Poisson distribution. Note that a univariate Poisson distribution may be inadequate to model it in 2D or 3D (as in deep water), but could provide valuable information on precise location of radioactive sources assuming that the detector (device) movement is nearly linear toward a source, or away from it. Least-square-based error minimization techniques are used when multiple sensors are used to simultaneously search for multiple sources (Knoll (2010) [47], Zhao, Zhang, and Sullivan (2019) [107]).

Problem 6.6   Suppose you are working for a company. You receive 5 internal email messages per hour and 12 external emails per working day (say 8 hr). What is the probability that you will receive (i) 30 or more emails per day, (ii) 10 or less internal emails in an hour, and (iii) no email messages in 5 mins.

Problem 6.7   If $X \sim \text{POIS}(\lambda)$, check whether $E(X^n) = \lambda E((X + 1)^{n-1})$.

Problem 6.8   If $X \sim \text{POIS}(\lambda)$ find the expected values (i) $E[\cos(\theta x)]$ and (ii) $E[\sin(\theta x)]$ using De Moivres theorem $\exp(i\theta) = \cos(\theta) + i\sin(\theta)$.

Example 6.9   Distribution of $X_1|X_1 + X_2 = n$ If $X_1 \sim \text{POIS}(\lambda_1)$ and $X_2 \sim \text{POIS}(\lambda_2)$ are independent, then the distribution of $X_1|(X_1 + X_2 = n)$ is $\text{BINO}(n, \lambda_1/(\lambda_1 + \lambda_2))$.

Solution: Consider the conditional probability $P[X_1|(X_1 + X_2 = n)] = P[X_1 = x_1] \cap P[X_2 = n - x_1]/P(X_1 + X_2 = n)$. Substitute the density to get $e^{-\lambda_1}\lambda_1^{x_1}/x_1! * e^{-\lambda_2}\lambda_2^{n-x_1}/(n - x_1)!/e^{-(\lambda_1+\lambda_2)}(\lambda_1 + \lambda_2)^n/n!$. Canceling out common terms from

---

[5]Some of the modern radar systems decouple the activity as *search radar* and *track radar* to facilitate multiple target tracking.

the numerator and denominator, and writing $(\lambda_1 + \lambda_2)^n$ in the denominator as $(\lambda_1 + \lambda_2)^{x_1} * (\lambda_1 + \lambda_2)^{n-x_1}$ this becomes

$$n!/[x_1!(n-x_1)!]\,(\lambda_1/(\lambda_1 + \lambda_2))^{x_1}\,(\lambda_2/(\lambda_1 + \lambda_2))^{n-x_1} \tag{6.19}$$

which is the binomial PMF with probability of success $p = \lambda_1/(\lambda_1 + \lambda_2)$. When $\lambda_1 = \lambda_2$ this becomes BINO$(n, 1/2)$.

Example 6.10   Poisson probabilities   If $X \sim \text{POIS}(\lambda)$, find (i) $P[X \text{ is even}]$ and (ii)$P[X \text{ is odd}]$.

Solution: (i) $P[X \text{ is even}] = e^{-\lambda}\left[\lambda^0/0! + \lambda^2/2! + \cdots\right]$. To evaluate this sum, consider the expansion of $\cosh(x) = [1 + x^2/2! + x^4/4! + \cdots]$. The above sum in the square-bracket is then $\cosh(\lambda) = \frac{1}{2}(e^\lambda + e^{-\lambda})$. From this we get the required probability as $e^{-\lambda} * \frac{1}{2}(e^\lambda + e^{-\lambda}) = \frac{1}{2}(1 + e^{-2\lambda})$ (ii) $P[X \text{ is odd}] = 1 - P[X \text{ is even}] = 1 - \frac{1}{2}(1 + e^{-2\lambda}) = \frac{1}{2}(1 - e^{-2\lambda})$.

Example 6.11   Conditional distribution   If $X \sim \text{POIS}(\lambda)$, find the conditional distribution of the random variable (i) $X|X$ is even and (ii) $X|X$ is Odd.

Solution: Let Y denote the random variable obtained by conditioning X to even values, and Z denote the random variable obtained by conditioning X to odd values. As the Poisson variate takes values $x = 0, 1, 2, \ldots$ the variate Y takes the values $Y = 0, 2, 4, 6, \ldots \infty$, and Z takes the values $Y = 1, 3, 5, \ldots \infty$. Using above example,

$$\sum_{i=0,2,4\ldots}^{\infty} f(y) - \sum_{i=1,3,5,\ldots}^{\infty} g(z) = \sum_{i \text{ even}} e^{-\lambda}\lambda^i/i! - \sum_{i \text{ odd}} e^{-\lambda}\lambda^i/i!$$

$$= \sum_{k=0}^{\infty}(-1)^k e^{-\lambda}\lambda^k/k! = e^{-2\lambda}. \tag{6.20}$$

From conditional probability, $P[X = k|X \text{ is even}] = P[X = k \cap X \text{ is even}]/P[X \text{ is even}] = f(y) = 2e^{-\lambda}\lambda^y/[y!(1 + e^{-2\lambda})]$ This gives

$$f(y; \lambda) = \begin{cases} 2e^{-\lambda}\lambda^y/[y!(1 + e^{-2\lambda})] \\ 0 \qquad\qquad\qquad\qquad \text{otherwise.} \end{cases}$$

Proceed exactly as above to get the PMF of z as $P[X = k|X \text{ is odd}] = P[X = k \cap X \text{ is odd}]/P[X \text{ is odd}]$ as

$$g(z; \lambda) = \begin{cases} 2e^{-\lambda}\lambda^z/[z!(1 - e^{-2\lambda})] \\ 0 \qquad\qquad\qquad\qquad \text{otherwise.} \end{cases}$$

6.7 RELATION TO OTHER DISTRIBUTIONS

The tail probabilities of a Poisson distribution is related to the incomplete gamma function as follows.

Theorem 6.12 *Prove that the survival probabilities of* $POIS(\lambda)$ *is related to incomplete gamma function as*

$$F(r) = P[x > r] = \sum_{x=r+1}^{\infty} e^{-\lambda} \lambda^x / x = \frac{1}{\Gamma(r+1)} \int_0^\lambda e^{-x} x^r \, dx. \tag{6.21}$$

Proof. Consider $\int_\lambda^\infty e^{-x} x^r \, dx$. Put $y = x - \lambda$, so that the range becomes 0 to ∞, and we get

$$\int_\lambda^\infty e^{-x} x^r dx = \int_0^\infty e^{-(y+\lambda)} (y+\lambda)^r dy = e^{-\lambda} \int_0^\infty e^{-y} \sum_{j=0}^{r} \binom{r}{j} y^j \lambda^{r-j} \, dy. \tag{6.22}$$

Take constants outside the integral to get $e^{-\lambda} \sum_{j=0}^{r} \binom{r}{j} \lambda^{r-j} \int_0^\infty e^{-y} y^j \, dy$. Put $\int_0^\infty e^{-y} y^j dy = \Gamma(j+1) = j!$ in (6.22), and expand $\binom{r}{j} = r!/[j!(r-j)!]$. The $j!$ cancels out giving

$$\int_\lambda^\infty e^{-x} x^r dx = e^{-\lambda} \sum_{j=0}^{r} \lambda^{r-j} \, r!/(r-j)!. \tag{6.23}$$

Divide both sides by $r!$, and write $r!$ as $\Gamma(r+1)$:

$$\frac{1}{\Gamma(r+1)} \int_\lambda^\infty e^{-x} x^r \, dx = \sum_{j=0}^{r} e^{-\lambda} \lambda^{r-j} /(r-j)!. \tag{6.24}$$

Put $r - j = k$ on the RHS. When $j = 0, k = r$ and when $j = r, k = 0$. Thus, the sum is equivalent to $\sum_{k=0}^{r} e^{-\lambda} \lambda^k / k!$. Subtract both sides from 1. The LHS is then $\frac{1}{\Gamma(r+1)} \int_0^\lambda e^{-x} x^r \, dx$. The RHS is $\sum_{k=r+1}^{\infty} e^{-\lambda} \lambda^k / k!$. This shows that the left-tail area of the gamma function is the survival probability of Poisson distribution. This proves our result. □

Problem 6.13 Prove that $\sum_{k=0}^{n} \exp(-\lambda) \lambda^k / k! = 1/n! \int_\lambda^\infty \exp(-x) x^n \, dx$.

The parameter λ of a Poisson distribution is assumed to be a constant in majority of applications. But the prior distribution of it may be known in some fields. Popular prior distribution assumptions are beta, exponential, and normal laws. Thus, the λ is assumed to be exponentially distributed in transportation engineering (in long-term average collision rate modeling on highway intersections or bridges).

Theorem 6.14 *Prove that the survival function of a central chi-square distribution with even degrees of freedom is a Poisson sum as*

$$1 - F_n(c) = \int_c^\infty e^{-x/2} x^{\frac{n}{2}-1} / 2^{n/2} \Gamma(n/2) \, dx = \sum_{x=0}^{n/2-1} \frac{e^{-\lambda} \lambda^x}{x!}, \quad \text{where } \lambda = \frac{c}{2}.$$

Proof. The proof follows easily because the χ^2 and gamma distributions are related as $\chi_n^2 \equiv$ GAMMA$(n/2, 1/2)$.

$$F_n(x) = 1 - e^{-x/2} \sum_{i=0}^{(n-2)/2} (x/2)^i / i!. \tag{6.25}$$

Putting $n = 2m$, we find that the CDF of central χ^2 with even DF can be expressed as a sum of Poisson probabilities.

$$F_{2m}(x) = \sum_{j=m}^\infty P_j(x/2), \quad \text{and} \quad \overline{F}_{2m}(x) = S_{2m}(x) = \sum_{j=0}^{m-1} P_j(x/2), \quad m = 1, 2, 3, \ldots, \tag{6.26}$$

where $P_j(u) = e^{-u} u^j / j!$. Sometimes the "n" (number of occurrences of an event of interest) may be known in advance and we may have to find that value of p for which the CDF $F(x; n, p) = \alpha$. This could of course be found by consulting a binomial table. An alternate solution is to use the relationship between binomial distribution and incomplete beta distribution, F distribution or χ^2 distribution. If $(1 - \alpha)$th quantile of an F distribution with $2(x + 1), 2(n - x)$ DF for numerator and denominator, respectively, is q, then the unknown p is given by $q/(q + (n - x)/(x + 1))$. Similarly, a confidence interval (CI) for the Poisson parameter λ can be found in terms of χ^2 distribution as

$$0.5 \chi^2_{1-\alpha/2;2r} \leq \lambda \leq 0.5 \chi^2_{\alpha/2;2(r+1)}. \tag{6.27}$$

□

Example 6.15 SF of Poisson distribution   Prove that the Poisson survival function is bounded by $(1 - \lambda/(x + 1)) f(x; \lambda)$.

Solution: By definition SF $S(x) = \sum_{x=k}^\infty \exp(-\lambda) \lambda^x / x!$ (we sum from k onward instead of $k + 1$). Write this in expanded form

$$\exp(-\lambda) \lambda^x / x! + \exp(-\lambda) \lambda^{x+1} / (x+1)! + \exp(-\lambda) \lambda^{x+2} / (x+2)! + \cdots. \tag{6.28}$$

Take $\exp(-\lambda)\lambda^x/x!$ as a common factor to get $\exp(-\lambda)\lambda^x/x![1 + \lambda/(x+1) + \lambda^2/(x+1)(x+2) + \cdots]$. As $\lambda^2/(x+1)(x+2) < \lambda^2/(x+1)^2$, and $\lambda^2/(x+1)(x+2)(x+3) < \lambda^2/(x+1)^3$, and so on, we get the inequality $S(x) < \exp(-\lambda)\lambda^x/x![1 + \lambda/(x+1) + \lambda^2/(x+1)^2 + \cdots]$. It is easy to see that the terms in the square brackets is $(1 - \lambda/(x+1))^{-1}$. Thus, $S(x) < \exp(-\lambda)\lambda^x/x![(1 - \lambda/(x+1))]^{-1} = [(1 - \lambda/(x+1))]^{-1} f(x;\lambda)$.

A relationship exists between a Poisson process and the exponential distribution. The distribution of inter-arrival times (distances in the case of 1D spatial processes) is exponentially distributed with parameter $1/\lambda$.

6.8   ALGORITHMS FOR POISSON DISTRIBUTION

Individual probabilities can be calculated using the forward recurrence

$$f_x(k+1;\lambda) = [\lambda/(k+1)] f_x(k;\lambda), \quad \text{with} \quad f_x(0;\lambda) = e^{-\lambda}. \tag{6.29}$$

When λ is large, $e^{-\lambda}$ is too small. This may result in loss of precision, or even underflow (in computer memory). As subsequent terms are calculated using the first term, error may propagate throughout the subsequent computation steps. A solution is to use the log-recursive algorithm suggested in Chattamvelli (1995) [14]. Another possibility is an algorithm that starts with the mode of the Poisson distribution, which then iteratively calculates subsequent values leftward (reverse) and rightward (forward).

Problem 6.16   If $X \sim \text{POI}(\lambda)$, show that the lines $y = x\Pr(x)/\Pr(x-1)$ and the line $y = mx + c$ intersect if $c = \lambda$ and $m = 0$.

The left-tail probabilities (CDF) $F_c(\lambda) = \sum_{j=0}^{c} P(j)$ converges rapidly for small λ values (see Figure 6.1). The CDF can be evaluated efficiently using $F_c(\lambda) = \gamma(c+1,\lambda)/\Gamma(c+1)$, where $\gamma(c+1,\lambda) = \int_{\lambda}^{\infty} e^{-t}t^c dt$ is the incomplete gamma integral.

Example 6.17   MD of Poisson distribution   Find the MD of Poisson distribution using the Power method in Chapter 1.

Solution: We know that the mean of Poisson distribution is λ. Using the theorem in Chapter 1, the MD is given by

$$\text{MD} = 2 \sum_{x=ll}^{\lfloor \mu \rfloor} F(x) = 2 \sum_{x=0}^{c} \gamma(x+1,\lambda)/\Gamma(x+1), \tag{6.30}$$

where $c = \lfloor \lambda \rfloor$, with proper correction term mentioned in Section 7.1 when λ is non-integer. The PGF of Poisson distribution $P(\lambda)$ is $e^{\lambda(t-1)}$. According to Theorem 6.12, the MDGF is the coefficient of $t^{\lfloor \lambda \rfloor}$ in the power series expansion of $2e^{\lambda(t-1)}/(1-t)^2$. Expanding the denominator as a power series and collecting the coefficient of $t^{\lfloor \lambda \rfloor - 1}$ gives

$$\text{MD} = 2e^{-\lambda} \sum_{i=0}^{k} (k-i)\lambda^i / i! \text{ where } k = \lfloor \lambda \rfloor. \tag{6.31}$$

6.9   APPROXIMATIONS

The Poisson distribution provides a good approximation to the Binomial distribution $B(n, p)$ when p is small, provided $\lambda = np > 10$, and n is large enough. The accuracy of this approximation increases as p tends to zero. As mentioned above, this limiting behavior is more dependent on the rate at which $p \to 0$ faster than $n \to \infty$.

As the variance of the distribution is λ, normal approximations are not applicable for small λ values. But when $\lambda \to \infty$, the variate $Y = (X - \lambda)/\sqrt{\lambda}$ is approximately normal. The continuity correction can be applied as before to get $P[(X; \lambda) \lesseqgtr k] = \Pr[Z \lesseqgtr \frac{k - \lambda \pm 0.5}{\sqrt{\lambda}}]$. The square root transformation is a variance stabilizing transformation in general for every distribution for which the mean and variance are equal, and in particular for the Poisson distribution [67]. Many approximations have appeared in the literature based upon this observation. For example, the Anscombe [2] approximation uses $2\sqrt{X + 3/8} \sim N(0, 1)$. An improvement to this is the $\sqrt{X} + \sqrt{X + 1}$ transformation to normality suggested by Freeman and Tukey. As the Poisson left-tail areas are related to the χ^2 right-tail areas, the individual Poisson probabilities can be approximated using the χ^2 probabilities.

6.10   GENERALIZED POISSON DISTRIBUTIONS

The classical single-parameter Poisson distribution discussed above is a suitable choice only when the distribution of events is rare and random. In some applications like microbiology, spectroscopy, etc., the distribution of events may not be rare (e.g., the concentration of detrimental bacteria in food may be high, say 40 colony forming units per grid). We could subdivide the experimental region into finer grids to have a Poisson process realized in it. Alternately, use one of the generalized Poisson distributions. Extensive discussions can be found in Consul (1989) [20] and Johnson, Kemp, and Kotz (2005) [39].

6.10.1   INTERVENED POISSON DISTRIBUTION (IPD)

The intervened Poisson distribution ([75], [79], [82]), and incidence rate restricted Poisson distribution ([77], [96], [88], Shanmugam (2006) [80]), and size-biased incidence rate restricted Poisson distribution [80] are extensions. An intervened Poisson process is decomposed into the sum of a ZTP distribution with parameter θ, and a regular Poisson process with parameter $\rho\theta$. This arises in various fields like epidemics, manufacturing, quality control, signal processing, optics, astrophysics, etc. Suppose an epidemic like COVID-19 is spreading fast, and is modeled as $Y \sim \text{POIS}(\theta)$ [92]. Assume that a public health mandate comes up with mitigation strategies that changes θ to $\rho\theta$ where $\rho < 1$ indicates that the containment measure was effective in reducing the spread, and $\rho \geq 1$ indicates that it was ineffective (intervention measures are more effective when ρ is closer to zero). If Z denotes the random variable after the mandate is implemented, then $X = Y + Z$ denotes the total count given by $\Pr[X = x] = \sum_{k=0}^{x-1} \Pr[Y = x - k] \; \Pr[Z = k | Y = x - k]$. Substitute the corresponding PMF to get

$$f(x; \rho, \theta) = \exp(-\rho\theta) \left((1 + \rho)^x - \rho^x\right) \theta^x / [x!(\exp(\theta) - 1)]$$
$$\text{for } x = 1, 2, \ldots, \infty, \quad \rho, \theta \in (0, \infty),$$
(6.32)

which for $\rho = 0$ reduces to the ZTP distribution. Take ρ^x as a common factor to get the alternate form

$$f(x; \rho, \theta) = \exp(-\rho\theta) \left((1 + 1/\rho)^x - 1\right) (\rho\theta)^x / [x!(\exp(\theta) - 1)],$$
(6.33)

which for $\rho = 1$ reduces to $\exp(-\theta) (2^x - 1) \theta^x / [x!(\exp(\theta) - 1)]$. The mean is $\mu = \theta(\rho + \mu_t/\theta)$ where μ_t is the mean of ZTP discussed in page 125. The variance is $\sigma^2 = \mu - \exp(\theta) (\theta/(\exp(\theta) - 1))^2$. A test for $H_0 = \rho = 1$ against $H_1 = \rho < 1$ appears in Shanmugam (1992) [78]. The PGF is $P_x(t, \rho, \theta) = \exp(\rho\theta t)(\exp(\theta t) - 1)/[\exp(\rho\theta)(\exp(\theta) - 1)]$ from which the mean $\mu = \theta(1 + \rho + 1/(\exp(\theta) - 1))$ and variance $\sigma^2 = \mu - \theta^2 \exp(\theta)/(\exp(\theta) - 1)^2$.

6.10.2   SPINNED POISSON DISTRIBUTION (SPD)

Shanmugam (2011) [81] introduced spinned Poisson distribution (SPD) with PMF

$$f(x, \rho, \lambda) = \frac{1 + \rho x}{1 + \rho\lambda} \exp(-\lambda)\lambda^x / x!$$
(6.34)

$0 < \lambda < \infty, 0 \leq \rho < \infty$, for $x = 0, 1, \ldots, \infty$, which reduces to $\text{POIS}(\lambda)$ when $\rho = 0$. This has mean $\mu = (1 + \lambda) - 1/(1 + \rho\lambda)$, and variance $\sigma^2 = (\mu - \lambda)[\mu/(\rho\lambda) + \lambda]$.

6.11   POISSON PROCESS

There are several processes in everyday life that can be modeled as Poisson process. Examples include visitors to service stations, cars at car wash stations, emergency calls received at special numbers (like 911, 100), subscriptions at stock markets, etc. A Poisson process is a probabilistic model that describes the random occurrences of discrete events over time or is associated with space (like surface of the earth or part of the sky) that satisfy the following properties:

1. events occur independently in different time intervals or different sub-regions;

2. probability of an event occurring in a very short time interval; δt is proportional to its length δt (for temporal processes) and area of sub-region for point processes; and

3. two or more events do not occur simultaneously, and probability of more than one occurrence in a very short time interval is negligible.

Since the events are not simultaneous as per the third condition, we could always find a fine tuned sub-interval of a Poisson process that follow the Bernoulli process (either occurrence or non-occurrence of the event). When space (2D or higher dimensional) is the frame of reference, a grid of proper size is chosen as the unit. For example, an imaginary grid of proper size is overlaid on the surface of a plate when the microbial density on it is assumed to follow a Poisson process. For temporal Poisson processes, the occurrence probability is expressed in terms of time (considered as continuous) as

$$P(x) = \exp(-\lambda t)\,(\lambda t)^x / x!, \tag{6.35}$$

where λ is the average rate of occurrence and t is the time period. As λ is changed, the probability of seeing the number of occurrences of events in one interval also gets changed because it represents the "most likely" number of event occurrence. It is not an integer in most practical applications. All rarely occurring events cannot be modeled blindly as Poisson process. As an example, some of the medical conditions, especially in elderly people, like dizziness, migraine headaches, seizures, fractures, dental caries, etc., are approximately Poisson distributed for some people. But there are also people for whom these symptoms do not occur continuously for a block of days, and sometimes occur multiple times per day. Thus, the Poisson model may be patient specific (with different durations of occurrence). Superimposition of two Poisson processes is analogous to the convolution of two Poisson random variables.

Although an exponential distribution with parameter λ describes the inter-arrival times between two events of a Poisson process, it is not easy to check it empirically. A simple solution is as follows: fix a duration τ and fit a Poisson model $P(\lambda)$. Then double the duration to 2τ and check if $P(2\lambda)$ is an appropriate fit.

It is called a spatial Poisson process when the events occur in a spatial setting (discrete or continuous). For example, if the epi-center of an earthquake is considered as a point, we get a point Poisson process on the surface (either on land or underwater) of the earth. The occurrence of the events follow the Poisson law, meaning that they are rare and random. The average time between the occurrence of events is assumed to be known. The t is replaced by τ that denotes the grid size (in 2D) or grid volume (in 3D) for spatial and volumetric processes (these are similar to the pixel and voxel used in 2D and 3D image processing applications). Any closed and bounded region can be used in 3D processes as in searches for continuous gravitational waves from supernova remnants or black-hole merging; air samples in deep mines, microchip factories, or liquid samples in hydrology, etc.

If $X(t)$ denotes a Poisson process, the quantity $[X(t + \delta) - X(t)]/\delta$ where δ is a small constant is called a Poisson interval, and the derivative of $X(t)$ w.r.t. time is called Poisson impulse.

6.12   FITTING A POISSON DISTRIBUTION

The likelihood function for a sample of size n is

$$L(\lambda) = \prod_{i=1}^{n} \exp(-\lambda)\lambda^{x_i}/x_i! = \exp(-n\lambda)(\lambda)^{\sum x_i}/x_1!x_2!\ldots x_n!. \tag{6.36}$$

Take log (to base e), differentiate w.r.t. λ and equate to zero to get $\sum x_i/\lambda - n = 0$ from which $\hat{\lambda} = \sum x_i/n = \bar{x}$. As the MLE estimate of the unknown parameter is the sample mean, a Poisson distribution can be fitted using a sample of counts as follows. Multiply the count with respective frequencies (at $x = 0, 1, 2$, etc.) and divide by the total count to obtain the sample mean:

$$\bar{x} = \sum_{x=0}^{n} x * count[x]/N \quad \text{where} \quad N = \sum count[x]. \tag{6.37}$$

Replace the unknown parameter λ by the sample mean, and obtain the fitted model. Substitute the x values to get the expected counts. An astute reader will notice that the count corresponding to $x = 0$ does not contribute to the numerator (but it is considered in the denominator). Thus, the numerator is the same irrespective of the magnitude of $count[0]$. This means that the higher the value of $count[0]$, the lower is the estimated mean (rarity increases when $count[0]$ is large). This situation gives rise to better fit of a Poisson model.

Both binomial and Poisson distributions are used to fit count data. The binomial distribution is used when a sample of size n has already been selected and the occurrence or non-occurrence of events are counted. Then the mean and variance are found as described in Chapter 2. The mean of a binomial distribution (np) is always greater than the variance (npq). The Poisson distribution does not pre-select a sample as in the case of binomial distribution. Here we count the total number of occurrences of an event of interest in a specific time interval. If the time interval is considered to be unity, the parameter is λ which is the total number of occurrences. If the time interval is assumed to be t, the parameter is λt. For spacial processes, t is replaced by either the area or volume of a uniform grid within which the counts are observed. In such cases, either the grid size is assumed to be unity or the "t" in the above is replaced by the area or volume of the grid. As the mean and variance are both the same, it is called an *equi-dispersion* property.

6.13 TRUNCATED POISSON DISTRIBUTION

Truncated distributions are obtained by omitting probabilities from the left, right or both tails. The simplest truncated distribution is the one obtained by omitting the zero-class. The ZTP, also called positive Poisson (PP) [79] distribution has PMF

$$f(x; \lambda) = \exp(-\lambda)\lambda^x/[(1 - \exp(-\lambda))x!], \text{ for } x = 1, 2, 3, \dots, \infty. \quad (6.38)$$

This is also called the *misrecorded* Poisson distribution. Divide numerator and denominator by $\exp(-\lambda)$ to get the alternate form

$$f(x; \lambda) = \lambda^x/[(\exp(\lambda) - 1)x!] = (\exp(\lambda) - 1)^{-1}\lambda^x/x!, \text{ for } x = 1, 2, 3, \dots, \infty. \quad (6.39)$$

The MGF is $[\exp(\lambda(\exp(t) - 1)) - \exp(-\lambda)]/(1 - \exp(-\lambda))$ with mean $\mu = \lambda/(1 - \exp(-\lambda))$. The variance is $\sigma^2 = \lambda/(1 - \exp(-\lambda))[1 - \lambda/(\exp(\lambda) - 1)] = \mu[1 - \mu \exp(-\lambda)]$, so that the ratio $\sigma^2/\mu = 1 - \lambda/(\exp(\lambda) - 1)$. The sum of n IID ZTP distribution has MGF $[\exp(\lambda(\exp(t) - 1))/(1 - \exp(-\lambda))]^n$. The corresponding PMF is

$$f(x; \lambda, n) = n!/(\exp(\lambda) - 1)^n \ S(x, n)\lambda^x/x!, \text{ for } x = n, n + 1, n + 2, \dots, \infty, \quad (6.40)$$

where $S(x, n)$ is Stirling number of the second kind (Tate and Goen (1958)[101]. This has mean $n\lambda/(1 - \exp(-\lambda))$ and variance $\mu(1 + \lambda - \mu/n)$. As $\sum_{x=0}^{n} f(x; \lambda) + \sum_{x>n} f(x; \lambda) = 1$, the above can be alternately expressed as $g(x; \lambda, n) = f(x; \lambda)/\sum_{x>n} f(x; \lambda)$, for $x = n + 1, n + 2, \dots$.

The rth rising inverse factorial moment is given by $E(1/(X + 1)^{[r]}) =$

$$1/(1 - \exp(-\lambda)) \sum_{x=1}^{\infty} \exp(-\lambda)\lambda^x/(x + r)!$$

$$= 1/[(1 - \exp(-\lambda))\lambda^r][1 - \exp(-\lambda) \sum_{j=0}^{r} \lambda^j/j!] = \gamma(r + 1, \lambda)/\Gamma(r + 1), \quad (6.41)$$

where $\gamma(r, \lambda) = \int_{\lambda}^{\infty} \exp{-y} y^{r-1} dy$. If $X_1, X_2, \dots, X_n$ are IID random variables, Singh (1978) obtained a necessary and sufficient condition for these to have ZTP distributions as $\sum_{j=1}^{k} X_j$ has a uniform multinomial distribution for each $k < n$.

6.14 ZERO-INFLATED POISSON DISTRIBUTION

In some application areas, the zero-count is either unusually high or unobserved. Consider the data about auto-accidents, or fire-accident of buildings. As majority of drivers are involved in accidents and majority of buildings do not have major fires, this data may be greater than expected. The zero-inflated Poisson (ZIP) model can capture excess zeros observed, and improve model fit.

$$f(x;\theta,\mu) = \begin{cases} 1 - \theta & \text{for} \quad x = 0; \\ \theta g(x;\theta,\mu) & \text{for} \quad x > 0, \end{cases}$$

where $0 < \theta < 1$ is the proportion of non-zero values and $g(x;\theta,\mu) = \exp(-\lambda)\lambda^x/[(1 - \exp(-\lambda))x!]$ is the PMF of ZTP distribution. This is called the ZIP model. This has MGF $M_x(t,\theta,\lambda) = (1 - \theta) + \theta \exp(-\lambda)(\exp(\lambda)t - 1)/(1 - \exp(-\lambda))$. From this the mean and variance follows as $\mu = \theta\lambda/(1 - \exp(-\lambda))$ and $\sigma^2 = \theta\lambda/(1 - \exp(-\lambda))[\lambda + 1 - \theta\lambda/(1 - \exp(-\lambda))]$.

6.15   DOUBLE POISSON DISTRIBUTION

The double Poisson distribution is an exponentiated combination of two Poisson probabilities with PMF

$$f(x,\lambda,\theta) = C(\lambda,\theta)\exp(-\lambda)\lambda^x/x. \tag{6.42}$$

6.16   APPLICATIONS

The Poisson distribution has been applied to various problems involving high uncertainty (low probability of occurrence) [11], [80]. Examples are the number of false fire alarms in a building, number of flaws in a sheet roll of newly manufactured fabric, number of phone calls received by a telephone operator in a fixed time interval, number of natural calamities like earthquakes, tsunamis, etc., in a fixed time interval (say one month), number of epidemics in a locality, number of deaths due to a rare disease, etc.[6] As the counts are converted into corresponding probabilities (relative frequencies), its magnitude does not matter in a Poisson model.

6.16.1   HEALTHCARE

Poisson distribution and its truncated version finds applications in various medical fields. As it can be used to model rare events, it is used in almost every medical field where rare diseases are to be modeled. One simple example is in emergency departments (given below). Other examples are in rare genetic disorders (due to genetic mutations), rare deaths, rare conditions, birth defects etc. Consider an example from toxicology. A toxin can enter the body through three primary pathways—ingestion through mouth (food and drink), inhalation (lungs), and contact with skin, mucous membranes, or eyes. If the total deaths due to toxicity follow a Poisson law, we could further model primary pathways using decomposition of the fitted Poisson model because majority of patients get toxicity through just one of the pathways.

Example 6.18   Emergency patients   An emergency department has noticed from past data that on the average 4 patients arrive every hour during day time. If 75% among the patients are

---

[6]Deaths due to anthrax during 1875–1895 was fitted by Bowley in 1901.

65 years or older, what is the probability of getting 5 or more patients 65 years or older in the next hour?

Solution: Let X denote the number of patients and Y denote patients 65 years or older. Then $P(x) = \exp(-4)4^x/x!$ and $P(Y|X) = 3/4$. Probability of getting 5 or more patients $= \sum_{x \geq 5} \exp(-4)4^x/x! = 1 - \sum_{x=0}^{4} \exp(-4)4^x/x! = 0.43347$. Assuming that patient arrivals are independent, the probability that all of them are 65 years or older is $(3/4)^5 = 0.2373$. From this the required probability is obtained as $0.2373 * 0.43347 = 0.102864$.

There are several problems in astronomy, biology, geology, ecology, and various engineering fields that can be modeled as random points in planar regions. As examples, suppose a patch of the sky (say one square inch) is examined for a special or rare object (like a supernovae, high luminosity star, spiral galaxy, etc.). The area of interest is first divided into uniform regions (called grid analysis). Then occurrences of the object of interest are counted. The probability of finding more and more objects will of course decrease in the case of Poisson distribution. The number of flaws or manufacturing engineering defects in fiber optics cable are assumed to follow a Poisson process with an average specified per length (say 100 ft) on sheets. The average number of NP photons per pulse through a fiber cable also is modeled using a Poisson process.

The area or length should be so chosen that zero occurrences are more than higher occurrences. If an event occurs multiple times in the selected area, we subdivide it into smaller areas until the condition is satisfied. If the area is so small that at most one event can occur in it, we could model it using binomial distribution. In this case, the probability of occurrence is given by p and n denotes the number of sub-areas under consideration.

Example 6.19 Structural damage A dam is built to withstand water pressure and mild tremors. Let X denote the number of damages resulting from a major quake. If X is distributed as POIS(.008), find the following probabilities: (i) probability of no damage, (ii) probability of at least 2 damages, and (iii) probability of at most 1 damage.

Solution: The PMF is $f(x, \lambda) = e^{-0.008}(0.008)^x/x!$, for $x = 0, 1, 2, \ldots$. Answer to (i) is $p_0 = e^{-0.008}(0.008)^0/0! = e^{-0.008} = 0.9920$. Answer to (ii) is $1 - P(0) - P(1) = 1 - 0.992032 - 0.007936 = 1 - 0.99996817 = 3.18298E - 05$. (iii) Probability of at most 1 damage $= \sum_{x=0}^{1} e^{-0.008}(0.008)^x/x! = 0.999968$.

Example 6.20 Defectives in shipment In a large shipment of articles, 0.2% are known to be defective. If 10,000 items are shipped find the probability that 10 or more are defective.

Solution: This can be solved by binomial distribution. But as $p = 0.20/100 = 0.002$ is small and $n = 10000$ is large, we could approximate it by a Poisson distribution with $\lambda = np = .002 *$

$10000 = 20$. Probability of 10 or more defectives is the tail probability of Poisson distribution for $x \geq 10$. This can be approximated using the gamma function.

6.16.2   ELECTRONICS

Semiconductor industry uses different methods to detect defects (physical defects due to foreign bodies or particles on chip surfaces, pattern defects, or systematic defects due to conditions of the mask or exposure process) on wafers. Defects are found using various techniques like dark field inspection, bright field inspection, scanning electron microscopy, and molten KOH method. Particles that get attached to the wafer usually result in random defects, while systematic defects occur at the same position on the circuit pattern of all the dies projected.[7] Almost 75% of the yield loss in IC fabrication is caused by random defects. Physical defects are assumed to follow the spatial Poisson law, which are modeled using grid-based methods (called wafer maps). Process engineers use wafer maps for yield enhancement and process improvements. If the entire die surface is taken as a unit and the mean number of defects is λ, the probability of finding x defects is given by the Poisson PMF. However, in practice, the yield predicted by the Poisson model is an under-estimate due to the fact that faults are spatially correlated on the die, and sometimes tends to cluster, which could indicate a process problem (for which compound Poisson distribution is more appropriate). Larger wafers tend to have spatial inhomogeneity. Perilous defects are those that make the chip totally useless.

Spatial defect pattern signatures at specific locations can throw insight into root causes in the IC fabrication process.

If the defects are counted in sub-regions of a chip, say one square mm area, the PMF of defects become $POIS(\lambda A)$ where A is the area of the die in mm^2 with PMF $\exp(-\lambda A)(\lambda A)^x/x!$. Put $x = 0$ to get the Poisson yield model as $y = \exp(-\lambda A)$ (from which $\lambda = -\ln(y)/A$). As the Poisson distribution is additive, the overall defect density can be decomposed into defects at different mask levels as $\lambda = \lambda_1 + \lambda_2 + \cdots + \lambda_n$, from which $y = \prod \exp(-\lambda_i A)$. This helps to understand yield improvement by reducing defect density.

Another application is in imaging and defect detection using x-rays, MRI, and ultrasound. Photon (quantum) noise produced by imaging equipments are assumed to follow the Poisson law, whereas electronic noise is assumed to be Gaussian distributed. A fundamental form of uncertainty associated with measurement in the above systems is called the Poisson noise. An x-ray source is assumed to transmit N photons at a specified energy level to a specified point on the detector or image receptor (called a pixel).[8] Detectors are either film-based sensors (onto which picture of x-rayed objects is captured) that rely on photo-sensitive chemical reactions, or digital sensors (in which photons are converted into electrons using photoelectric effect). The $N \sim POIS(\lambda)$ and the number M of photons falling on a pixel of the detector is also assumed to

---

[7] Defects are identified by keeping wafers of the same pattern side by side to compare defect patterns.

[8] Energy of a single photon is $E = hf$ where h is Planck's constant and f is the frequency of corresponding wave.

be Poisson distributed. Chances are that some photons will pass unaffected through the object and fall on the detector in which case M and N are identical. Other photons that pass through tissues (bones, muscles, etc., in medical x-rays) is called faithful shadowgram as they are the primary rays that produce useful information for practitioners. Intensity of some photons that interact with the object as it passes through it will decrease as it reaches the detector according to the exponential law as $I = I_0 \exp(-\mu t)$ where μ is the attenuation coefficient, t is the thickness, and I_0 is a constant. In the extreme case it may be almost blocked out (reduced to near zero) when dense objects like some metals are x-rayed (but most metals are not opaque to x-rays). If we assume that photons interact independently, $\Pr[M = m|N = n] = \binom{n}{m} p^m (1 - p)^{n-m}$ where p is the survival probability of photons traveling through a fixed ray and is given by Beer's law as $p = \exp(- \int \mu(z)dz)$. Using the law of total probability, the unconditional distribution of M is Poisson (λp). Hence, the number of photons that hit the detector is also Poisson distributed. Although x-rays can reveal anatomical features of the object being x-rayed, modern medicine uses ingested tracers that emit γ-rays that can reveal physiological functioning of internal organs and body parts (like veins) in greater detail.

The reliability of stress-strength models used in electronics, mechanical engineering, civil engineering, and many other fields can be modeled using the Poisson law. Here stress denotes any component, agent, force, or entity that induces failure or malfunctioning and strength denotes those resisting failure. If the cycle-times are known to be random, the entire system reliability is expressed as $R(t) = \sum_{k=0}^{\infty} p_k(t) R_k(t)$ where $p_k(t)$ is probability of k cycles occurring in time $[0, t]$ and $R_k(t)$ is probability of k of the components working in this cycle. The number of cycles occurring in a fixed time interval is assumed as $\text{POIS}(\lambda t)$ where λ is the mean number of occurrences in unit time.

6.16.3   POISSON REGRESSION

A useful distribution in psychology, dentistry, and epidemiological studies is a Poisson distribution truncated at 0. It is also used in several engineering fields and in search engine optimization. Assume that a user query returns a large number of matches, that are displayed by a search engine in discrete screenfuls of say ten matches each. Then the number of pages viewed by a surfer can be modeled as a ZTP law or a zipf law. The PMF is given by

$$p_x(\lambda) = e^{-\lambda} \lambda^x / [(1 - e^{-\lambda})x!] = \lambda^x / [(e^{\lambda} - 1)x!], \quad x = 1, 2, \ldots, \tag{6.43}$$

where the second expression is obtained from the first by multiplying the numerator and denominator by e^{λ}. The mean and variance are $\lambda/(1 - e^{-\lambda})$. This model is used in ZTP regression where the outcomes (counts) are rare. Suppose the data generated by a Poisson process is further sub-divided using other variables (like gender, age, or income groups, etc.):

$$\log(\lambda) = b_0 + b_1 x_1 + b_2 x_2 + \cdots + b_p x_p. \tag{6.44}$$

Such models are called Poisson regression models. It is used to explain how a set of two or more explanatory variables contribute to the observed count data. It finds extensive applications in travel and transportation industry, vital statistics, actuarial sciences, medical, and allied sciences. Ingredients and byproducts of several chemical and pharmaceutical industries are known carcinogens. Assume that a single exposure of such chemicals for a prolonged time period t (say 1 year) can result in cancer for some people. If d denotes the dose (amount of chemical that enters human body through food, water, air or through skin in period T and $Pr(t)$ denotes the probability that the individual will have cancer, then $Pr(t) = 1 - \exp(-(c_0 + c_1 d))$ is a Poisson regression model where c_0 and c_1 are constants to be determined from a sample (say using least squares principle). This can also be written as $\log(1 - Pr(t)) = -(c_0 + c_1 d)$. Putting dosage $(d = 0)$ should give us the background cancer incidence rate (BCR) (persons who are not exposed to the chemical getting cancer). If BCR is small, the above can be written as $Pr(0) = 1 - \exp(-c_0)$. Expand using $\exp(-x) = 1 - x/1! + x^2/2! - \cdots$ to get the first two approximants as $P(0) = c_0$ and $Pr(0) = c_0(1 - c_0/2)$. If the dosage level d is small (as when the carcinogen enters the body through air or skin), the above becomes $Pr(d) = 1 - [1 - (c_0 + c_1 d)]+$ terms of higher order $\simeq Pr(0) + c_1 d$. Subtract the background risk to get the additional risk due to the carcinogen as $c_1 d$. An extension of the above is used in tumor formations, and many opportunistic diseases modeling that depend on multiple factors as

$$Pr(d) = 1 - \exp(-(c_0 + c_1 d + c_2 d^2 + \cdots + c_n d^n)), \qquad (6.45)$$

where c_i's are constants to be determined from available data. The same model can be modified for occupational exposure to cancerous sources like x-rays, radioactive materials, petrochemicals, plastics, asbestos factories, etc. It is also used by the travel industry to estimate *trip generation* estimates as

$$Pr(t_k) = \exp(-\lambda_k)\lambda_k^{t_k}/t_k!, \qquad (6.46)$$

where λ_k varies among customers (or households), and denotes the expected number of trips (shopping trips, vacation trips, sight-seeing trips, business trips, etc.) in a time period t_k and is itself a random variable having a known form of relationship (log-linear relation is usually assumed in travel industry as $\lambda_k = \exp(-\sum_j b_j y_j)$) household variables contributing to trip planning or generation. Suppose $x_1, x_2, \ldots, x_p$ are p regression variables that jointly influence a count response variable Y that follows the Poisson distribution. Then $\lambda' = E(Y|X) = b_0 + \sum_{j=1}^{p} b_j y_j$ is a linear model. As the Poisson mean is always positive, we assume $\lambda' = \log(\lambda) = b_0 + \sum_{j=1}^{p} b_j y_j$ as our model. The coefficients b_j are determined using least squares method.

6.16.4   MANAGEMENT SCIENCES

Customers arrive in person (as in post offices) or over web or networks (like telephone) in various businesses. Poisson models are used whenever the arrivals are random and independent of each other. The probable maximum number of customer arrivals can be computed after adjustment for periodic variations so that staff requirements can exactly be predicted. This will reduce customer churns and increase business profits. Next consider an application of the Poisson distribution to businesses that loan out goods and services to customers. The location of business and the time of loan are unimportant to us as they may be located anywhere on the world. This is especially true for web-based e-commerce businesses. A modeler may wish to have maximum profit, and minimum storage and related costs; in the case of businesses that require storage space or manpower to keep items for loan (like cars, tractors, boats, lawn mowers, pumps, books, etc.). Potential customers and customer goodwill are lost when inventory does not have enough items to loan out. This also applies to services that require manpower (as in onsite repair services, software consultancy services, etc.). Additional items or manpower can be retained (called standby redundancy) if the extra cost involved is justified by the demand pattern over time. This is used by engineers for optimal asset management decision making on an ongoing basis. If D denotes the expected demand, and S is the number of spares, the $\Pr[X \leq S]$ is given by the Poisson CDF. This is called the provisioning model in inventory control. A modified provisioning model is applied for machinery that require many spares (like automobiles, aircraft, etc.) as $P = \prod_{k=1}^{m} P(s, k)$ where $P(s, k)$ is the cumulative probability of having k spares for sth spare part, which again is Poisson distributed. A Poisson model can predict the probability of expected arrival of orders at specified trading venues, or over-subscription periods. It is also used in various fields of finance. As examples, it is used to model the number of crashes (and big bounces) in equity markets, bond defaults in portfolios, arrival of buy or sell orders, new transactions at specified trading venues, and number of product returns of a particular type and in algorithmic trading. Banking applications include bank frauds for very large amounts, cyber attacks on banks and financial institutions, large mergers and acquisitions, bank failures, etc. It is also used in jump-diffusion models, customer segmentation models (e.g., in terms of investment, dormant, or unclaimed amounts), etc.

6.16.5   QUANTUM CRYPTOGRAPHY

Some of the quantum cryptography algorithms use diode lasers that generate weak laser pulses to be sent through optical fiber cables. Hackers and intruders can easily intercept and manipulate the message if multiple photons are transmitted (in unit time). Although rarely achieved, messages are highly secure when single photons are used for communication in a given time interval. Laser attenuators are used for this purpose. A coherent photon state with a preset value for the mean λ (of the Poisson distribution) is used for this purpose. Then the conditional probability that a nonempty coherent laser pulse has more than one photon is

$$P(x > 1|x > 0, \lambda) = (1 - \exp(-\lambda) - \lambda \exp(-\lambda))/(1 - \exp(-\lambda)), \tag{6.47}$$

where λ is a small constant.

As Poisson distribution has support $[0, \infty]$, a transformation of the form $y = x^{1/k}$ for $k \geq 2$ can be used to achieve rapid normality. This is because the resulting PMF is given by $f(y) = 2yf(x; y^2)$. In particular, $y = \sqrt{x}$ is commonly used for rapid convergence to normality [67].

Example 6.21 Probability of failure   A large city has 75 transformers installed in their power distribution network. A transformer may fail due to spark fires, fuse failures, or extreme weather conditions (like up-rooting due to high winds, flooding, etc.). It is observed from past data that that 35 transformers fail on the average per year. Failures occur independently of each other and failed transformers are to be repaired immediately. The repair person takes one day to repair it including to-and-fro travel time. The power supply company has one regular repair person and on-demand repair persons. Find the probability that (i) on-demand repair persons may have to be called on a day by restoring failed transformers in 24 hrs (ii) given that some transformers have failed, what is the probability that on-demand repair persons may have to be called?

Solution: This problem can be solved using binomial or Poisson distribution. Poisson model is better as failures are rare events. (i) Probability for calling on-demand repair persons is the same as probability of two or more failures (as the regular person takes one day to repair). This is given by $1 - P[\text{no failure}] - P[\text{one failure}]$. As there are 35 failures in one year, we have $\lambda = 35/(75 * 365) = 0.0012785$ failures in 24 hours. Hence, the required probability is $1 - \exp(-\lambda) - \lambda \exp(-\lambda) = 0.0000008166$. In part (ii) we are given that one or more failures have occurred. Hence, we can model it as a ZTP distribution.

6.16.6   POISSON HETEROGENEITY TEST

This test aims to check if a given sample of size n has been drawn from a Poisson distribution. It is used in several experiments that have a spatial frame of reference. Consider a bacterial culture grown on a flat microscopic plate. Limiting dilution assays used in microbiology uses a chemical or a pathogen as an positive agent (like a catalyst used in chemical reactions) that allows concentration of an organism. The analyst can detect the presence or absence of these agents. A succession of dilutions (usually following a geometric progression) are then applied to obtain the correct degree of dilution needed to eliminate the effect of the agent. A count is accumulated for each narrow strip of equal area (say small square regions on the plate). Sustained growth of the organism may not happen due to a multitude of reasons, resulting in heterogeneity in the counts. Similarly, clustered growth could occur due to extraneous factors (like uneven plates, temperature or moisture variations, etc.). If distinct heterogeneous clusters are present, we could model the cluster distribution by one Poisson distribution, and microbe distribution in each cluster using another Poisson distribution with different grid sizes. Common difference ratios other than 1/2 are also used in practice. A similar technique called extrapolation from high to low doses is used in environmental engineering, pathology, cytology, and many other biological sciences. As an example, carcinogenic chemicals are tested on animals (mice, rabbits, monkeys) at high doses to ascertain the time period for cancer onset before it is tried at extremely low doses (diluted to 10^{-5} or less) on human subjects.

The Poisson heterogeneity test is appropriate in such situations. This test uses a test statistic $t = \sum(x_i - \bar{x})^2/\bar{x}^2$. It also finds applications in environmental science and engineering. Consider the population of a country classed into age groups (0–10), (10–20), (20–30), ..., > 100. This is approximately Poisson distributed in most countries, but a central-bulging occurs in some countries due to extraneous factors like governmental regulations, wars and famines, natural disasters, etc.

Problem 6.22 A cruise-ship has a capacity of 3,000 passengers. As there are several "no-show" at the last moment, the company decides to overbook passengers beyond 3,000 head count. If there are 5% "no-shows" on the average, what is the maximum excess count they can overbook so that booked passengers will not be bumped out. Hint: Take $n = 3000 + k$ where k is an integer that represents overbooking. None will be bumped if the total number of "no-show ups" is $\geq k$.

6.17 SUMMARY

This chapter introduced the Poisson distribution and its properties. Although a single-parameter Poisson distribution suffices for most of the modeling situations, there are several fields where it underfits the data. Examples are microbiology and spectroscopy. If the number of colony forming units (CFU) in a microscopic plate is large, and uniform, a finer grid size may be employed to fit a Poisson law. If it is large and non-uniform, the Poisson law is almost symmetric (because it theoretically tends to the normal law) whereas microbial distributions are often skewed to the right. One solution is to use a linear combination of Poisson variates. Alternately, the generalized Poisson distribution may be used as it has additional parameters. Some of the discrete exponential decay models can be approximated by a Poisson law. Examples are exponential resource production models (e.g., face masks production during COVID pandemic) used in manufacturing engineering. Exponential growth models encountered in microbiology, management, and many other fields cannot be modeled directly by a Poisson law. For instance, carbon emissions by developed countries and power generation by alternate sources (like solar and wind) in developing countries are growing at a steady rate. If the growth is viewed in reverse chronological manner, it could be approximated by a Poisson distribution. One good example is the growth of COVID-19 virus infection which is exponential in some countries. Other examples are power consumption in developing countries, and population growth in cities or countries.

CHAPTER 7

Hypergeometric Distribution

> **After finishing the chapter, readers will be able to ...**
>
> - Understand hypergeometric distribution.
>
> - Describe hypergeometric distribution and its properties.
>
> - Explain Capture-Mark-Recapture model.
>
> - Apply hypergeometric distribution in practical situations.

7.1 INTRODUCTION

The classical hypergeometric distribution was introduced by the French mathematician Abraham De Moivre[1] in 1711. This was applied to some practical problems by Cournot (1843) [23]. Suppose a researcher is interested in studying a group of N entities (people, places, planets, machines, living organisms, web sites, software programs, data packets, or human-built structures like dams, buildings, bridges, etc.) that either possess an attribute or does not. The attribute may be a condition, property, distinctive nature, characteristic, or feature. It is known from past data that p percent of the entities possess the attribute, so that there are Np entities with the attribute, and the remaining Nq entities that do not have it.[2] In other words, among the N items, $k = Np$ are of one kind, and the rest $(N - k) = Nq$ are of another kind. We assume that the two kinds are indistinguishable. Practical experiments that result in a hypergeometric distribution can be characterized by the following properties:

1. The experiment consists of sampling *without replacement* from a dichotomous population.

2. The trials can be repeated independently n times under identical conditions.

3. The probability of occurrence of the outcomes p_k varies (increases) from trial to trial until the experiment is over.

4. The random variable X denotes the number of times one of the dichotomous groups is selected.

---

[1]De Moivre obtained it as a solution to the urn problem proposed by Huygens.
[2]It is assumed here that $p + q = 1$, so that $Np + Nq = N$.

Suppose we sample n items *without replacement* from the set (called a hypergeometric trial). The number of items x of the first kind is then given by the PMF

$$f(x; k, N, n) = \binom{k}{x}\binom{N-k}{n-x}\Big/\binom{N}{n}, \quad \text{where} \quad x = 0, 1, 2, \ldots, \min(n, k). \tag{7.1}$$

This is called the hypergeometric distribution (HGD), which has three parameters k, N, and n. It is the distribution of x successes in n trials without replacement from N items in which there are k successes. This can be derived using the following argument. As there are k items of one kind, we can choose x items from it in $\binom{k}{x}$ ways. To make the count to n, we need to select further $n - x$ items. But these can be selected from $(N - k)$ items of second kind in $\binom{N-k}{n-x}$ ways. By the product rule, the probability of this happening is
(number of ways to select x of the first kind from k items) × (number of ways to select $n - x$ of the second kind from $N - k$ items)/(total number of ways to select n items of any type from N items).
If x_i denotes the ith sampled item, we may associate a Bernoulli variable as $x_i = 1$ if ith sampled item is of first kind, and $x_i = 0$ otherwise. Let $x = \sum_{i=1}^{n} x_i$. Then the probability that $x = i$ is given by the hypergeometric distribution. The difference between binomial and hypergeometric distributions is that the trials are independent in binomial distribution, whereas it is dependent (sampling is done without replacement, so that the odds change after each item is drawn) in hypergeometric distribution.

7.2   ALTERNATE FORMS

Another form that uses $k = Np$ can easily be obtained as

$$f(x; p, N, n) = \binom{Np}{x}\binom{N-Np}{n-x}\Big/\binom{N}{n} = \binom{Np}{x}\binom{Nq}{n-x}\Big/\binom{N}{n}, \tag{7.2}$$

where $x = 0, 1, 2, \ldots, \min(n, k)$. This is indeed a PMF, which could be proved from the identity $(1 + x)^{Np}(1 + x)^{Nq} = (1 + x)^{Np+Nq} = (1 + x)^N$. Expand each binomial term in (7.2), and use $(a/b)/(c/d) = ad/bc$ to get $f(x; p, N, n) =$

$$(Np)!/[x!(Np - x)!]\,(Nq)!/[(n - x)!(Nq - n + x)!]\,n!(N - n)!/N!. \tag{7.3}$$

Now rearrange to get many alternative forms, one of which is $f(x; p, N, n) =$

$$n!/[x!(n - x)!]\,[(Np)!(Nq)!/N!]\,(N - n)!/[(Np - x)!(Nq - n + x)!]$$

$$= \binom{n}{x}\binom{N-n}{Np-x}\Big/\binom{N}{Np}, \tag{7.4}$$

using $N - n - Np + x = Nq - n + x$. This could also be written in terms of gamma functions as $f(x; p, N, n) =$

$$\binom{n}{x}(N-n)!/N! \ \Gamma(Np+1)\Gamma(Nq+1)/[\Gamma(Np-x+1)\Gamma(Nq-n+x+1)]. \quad (7.5)$$

Expand each factorial, and cancel out common terms to get the alternate form

$$f(x; p, N, n) = \binom{n}{x}(Np)_{(x)}(Nq)_{(n-x)}/N_{(n)}, \quad (7.6)$$

where $N_{(r)} = N(N-1)\dots(N-r+1)$, etc. denotes Pochhammer falling factorial.

Write $(Nq)! = (Nq)(Nq-1)\dots(Nq-n+1)\ (Nq-n)!$ and $(Nq-n+x)! = (Nq-n+x)(Nq-n+x-1)\dots(Nq-n+1)\ (Nq-n)!$ in (7.3). Cancel out $(Nq-n)!$ to get another form as

$$f(x; p, N, n) = (Nq)_n/(N)_n \left[(Np)_{(x)}(n)_{(x)}/(Nq-n+x)_{(x)} \ 1/x!\right], \quad (7.7)$$

where expressions in the square braces have x as the variable in Pochhammer falling factorial. This is easily identified as the coefficient of t^x in the hypergeometric series $(Nq)_n/(N)_n \ {}_2F_1(-n, -Np; Nq-n+1, t)$, where

$${}_2F_1(a, b; c; x) = 1 + (ab/c)x/1! + a(a+1)b(b+1)/[c(c+1)] \ x^2/2! + \cdots. \quad (7.8)$$

is the Gaussian hypergeometric series [54].

As the expression involves binomial coefficients, there is a natural symmetry involved in the above PMF. Swapping the roles of p and q results in a dual hypergeometric distribution (DHGD). Instead of sampling x items from the first kind, we could take x items from the second kind and $(n-x)$ items from the first kind (defectives are assumed as "success" in manufacturing and SQC environments). This gives us the alternate PMF:

$$f(x; k, n, N) = \binom{N-k}{x}\binom{k}{n-x}/\binom{N}{n}. \quad (7.9)$$

This could also be written in terms of Np and Nq, which could be expressed in terms of gamma functions.

$$f(x; k, n, N) = \binom{Nq}{x}\binom{Np}{n-x}/\binom{N}{n}. \quad (7.10)$$

To impose the range for both these forms, we modify the range of x values as $0, 1, 2, \dots, \min(m, N-m, n)$. As all combination terms must exist, x varies between $\max(0, n+m-N)$ and $\min(m, N)$.

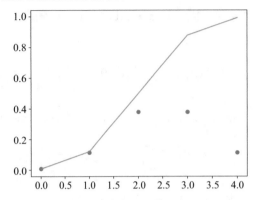

Figure 7.1: PMF and CDF of HGD(N=12, n=6, k=5).

7.3   RELATION TO OTHER DISTRIBUTIONS

If the number of items selected (n) is one, we have a Bernoulli distribution with $p = k/N$. Swapping the roles of success (type-1 entities) and failure (type-2 entities) results in a DHGD. For $n = N - 1$, the DHGD is Bernoulli distributed. If n is small and fixed, and N is relatively very large, the distribution is approximately BINO(n, p). Similarly, when N is very large, and $p = k/N$ remains a constant, the negative hypergeometric distribution tends to the negative binomial NBINO(n, p). When $p \to 0$ and $N \to \infty$ such that Np remains a constant, the hypergeometric distribution tends to the Poisson law POIS(λ) where $\lambda = nk/N$ is the mean. It tends to the normal law $N(\mu, \sigma^2)$ when N is large and $p \to 1/2$, where μ and σ^2 are given below. Equivalently, $(X - \mu)/\sigma$ tends to the standard normal law when N is large and p is not very small (preferably $p \to 1/2$). Relation to the geometric and negative binomial distributions are discussed in the application section below. Relation to the multinomial distribution is discussed in Chapter 10. As the PGF (and MGF) can be expressed in terms of hypergeometric functions, it is related to many other distributions that are related to hypergeometric functions [54]. The multivariate hypergeometric distribution is obtained when the population comprises of k groups instead of two in the classical hypergeometric distribution. This has PMF

$$f(x; \mathbf{M}) = \prod_{j=1}^{k} \binom{M_j}{x_j} / \binom{N}{n}, \tag{7.11}$$

where $\sum x_j = n$, M_j is the number of items in the jth class or group, and $\mathbf{M}$ is the vector of M_j values.

This distribution can be obtained from Polya's urn model as follows. Consider an urn containing two types of balls (say r red and w white). Balls are drawn one-by-one at random without looking at the color. Assume that after noting down the color of the selected ball, we return c *additional* number of balls of the same color to the urn, where c is an integer (positive or negative). Different distributions arise for different values of c. No *additional balls* are returned, but the selected ball is returned back for $c = 0$. Then, subsequent selections become *with re-placement* so that it results in binomial distribution $BINO(n, r/(r + w))$, if red balls denote success, or its dual binomial otherwise. Similarly, $c = -1$ implies that even the selected ball is not returned. This results in positive hypergeometric distribution. Successes and failures are contagious when c is positive. All other values of c results in Polya distribution. Instead of assuming c to be the *additional balls* are returned, we could assume c as the total balls returned. In this case, $c = 1$ results in binomial, $c = 0$ in positive hypergeometric distributions. Removing balls until a specified number k balls of a specific color is selected results in either negative binomial (with replacement) or negative hypergeometric (without replacement) distributions. Note that returning balls to the urn is a conceptual logic. It has many applications in practical situations. Consider the problem of modeling the spread of COVID-19 or some similar epidemic within a city where there are m confirmed cases. Each time active patients are sampled, they are kept in isolation (removed from the population), and replaced with n number of other persons with whom the removed patient might have come into close contact (they are likely exposed to the virus). Now the population size is increased by $(n - 1)$ for each such removed patient. Thus, the population size grows rapidly so that it can be accurately modeled. Similarly, not returning balls to the urn means that the occurrence of an event has an aftereffect in increasing the probability of sampling (as population size is reduced).

7.4   PROPERTIES OF HYPERGEOMETRIC DISTRIBUTION

This distribution has three parameters, all of which are integers (in the classical distribution). The PMF could become numerically intractable for some parameter values due to large numbers that may appear in the combination terms. Solutions exist to ease the computation [4], [29]. The recurrence relation for the PMF is

$$f(x + 1; k, n, N) = f(x; k, n, N) * \frac{(n - x)(k - x)}{(x + 1)(n - k + x + 1)}. \tag{7.12}$$

As the parameters are all related, the following symmetries follow easily (i) $f(x; k, n, N) = f(x; n, k, N)$, (ii) $f(x; k, n, N) = f(k - x; k, N - n, N)$, and (iii) $f(x; k, n, N) = f(n - x; N - k, n, N)$. Replace x by $x - 1$ in (7.1) to get

$$f(x; k, n, N) = f(x - 1; k, n, N) * \frac{(n - x + 1)(k - x + 1)}{x(N - k - n + x)}. \tag{7.13}$$

The CDF does not have closed form except for particular parameter values. But due to the symmetry relationships, it is possible to get $F(x; k, n, N - k) = 1 - F(n - x - 1; N - k, n, k) =$

$F(N - k - n + x; N - k, N - n, k)$. Several approximations for the CDF in terms of binomial CDF are available in the literature, the most notable among them by Moleenar (1973) [57], Lopez-Blazquez and Salamanca(2000) [52].

$$F(x; k, n, N - k) \simeq \sum_{j=0}^{x} \binom{n}{j} p^j (1 - p)^{n-j}, \tag{7.14}$$

where $p = (Np - x/2)/[N - (n-1)/2] - n(x - np - 0.5)/6[N - (n-1)/2]^2$, which could in turn be expressed in terms of incomplete beta function as $I_q(n - x, x + 1)$, where $q = 1 - p$. As the binomial tail probabilities can be expressed in terms of F distribution, and Poisson tail probabilities by a chi-square distribution, these may also be used for large n. The distribution of the number of success-runs in hypergeometric trials is itself hypergeometric (Godbole (1990) [27]).

7.4.1   FACTORIAL MOMENTS OF HYPERGEOMETRIC DISTRIBUTION

Factorial moments are easier to find due to the $x!$ in the denominator (of both the forms (7.1) and (7.2)). The rth falling factorial moment

$$\mu_{(r)} = E[x_{(r)}] = E[x(x-1)\ldots(x-r+1)]$$

$$= \sum_x x(x-1)\ldots(x-r+1) \binom{k}{x} \binom{N-k}{n-x} / \binom{N}{n}$$

$$= \left[1/\binom{N}{n} \right] \sum_{x=r}^{n} x(x-1)\ldots(x-r+1)(k)_x/x! \binom{N-k}{n-x}. \tag{7.15}$$

Cancel out $x(x - 1)\ldots(x - r + 1)$ from $x!$ in the denominator, and write $(k)_x = k(k-1)\ldots(k-r+1)(k)_{x-r}$, and take it outside the summation. This gives $\mu_{(r)} = [(k)_r/\binom{N}{n}] \sum_{x=r}^{n} \binom{k-r}{x-r}\binom{N-k}{n-x}$. Change the indexvar using $u = x - r$ to get

$$\mu_{(r)} = \left[(k)_r/\binom{N}{n} \right] \sum_{y=0}^{n-r} \binom{k-r}{y} \binom{N-k}{n-y-r}. \tag{7.16}$$

Using Vandermonde's identity, this becomes

$$\mu_{(r)} = (k)_r \binom{N-r}{n-r} / \binom{N}{n} = (Np)_r \binom{N-r}{n-r} / \binom{N}{n}, \tag{7.17}$$

so that the factorial MGF (FMGF) is

$$_2F_1[-k, -n, -N; -t]. \tag{7.18}$$

7.4.2   MEAN AND VARIANCE OF HYPERGEOMETRIC DISTRIBUTION

The mean is easily obtained from factorial moments given above by putting $r = 1$ as nk/N. Associate a Bernoulli variable x_i with each item of first kind as $x_i = 1$ if ith item of first kind is selected, and $x_i = 0$ otherwise. $E[x_i] = \Pr[x_i = 1] = k/N$, and $V[X_i] = k/N(1 - k/N)$. Let $X = \sum_{i=1}^n X_i$. Then $E[X] = E[X_1] + E[X_2] + \cdots + E[X_n] = nk/N = np$. Thus, the means of binomial and hypergeometric distributions are both np. From this an alternate representation can be obtained as

$$f(x; \mu, N, n) = \binom{N\mu/n}{x}\binom{N(1 - \mu/n)}{n - x}/\binom{N}{n}. \tag{7.19}$$

The variance of X is most easily obtained using factorial moments as $V(X) = E[X(X - 1)] + E(X) - E(X)^2$. From above, $E[X(X - 1)] = n(n - 1)\,k(k - 1)/[N(N - 1)]$. Substitute in the above to get the variance as $(nk/N)(1 - k/N)(N - n)/(N - 1)$, which is $n(N - n)/(N - 1)$ times $V[X_i]$. Put $k/N = p$ to get the variance as $npq\,(N - n)/(N - 1)$ showing that the variance is $(N - n)/(N - 1)$ times the variance of the binomial distribution. This is called the finite population correction factor (FPCF). As $n > 1$, the numerator of FPCF is always less than the denominator. This means that FPCF is always < 1 showing that the variance of hypergeometric distribution is always less than the corresponding binomial distribution. Divide the numerator and denominator of FPCF by N to get FPCF $= (1 - n/N)/(1 - 1/N)$. This shows that the FPCF $\to 1$ as $n \to 0$ for large N, and FPCF $\to 0$ as $n \to N$. Replace nk/N on the RHS by μ and write the multiplier as $(1 - k/N)*$FPCF. This shows that $\sigma^2 < \mu$ as both $(1 - k/N)$ and FPCF are fractions. From $f(x + 1; k, n, N) = f(x; k, n, N)* \frac{(n-x)(k-x)}{(x+1)(n-k+x+1)}$ we see that $f(x + 1; k, n, N)$ is greater than or less than $f(x; k, n, N)$ according as $\frac{(n-x)(k-x)}{(x+1)(n-k+x+1)}$ is $>$ or < 1. Thus, the probability increases steadily, reaching a maximum at $(n + 1)(k + 1)/(N + 2)$. Therefore the mode of the distribution is $M = \lfloor(k + 1)(n + 1)/(N + 2)\rfloor$, which is greater than the mean. It is unimodal if M is a fraction, and bimodal with modes at $M - 1$ and M if it is an integer.

Covariance (for $i < j$) is the probability that both i and j are successful for $i \neq j$ is given by $E(X_i X_j) - E[X_i]E[X_j]$. As the product of two IID BER(p) is also Bernoulli distributed (Chapter 2, page 22), $x_i x_j$ will be 1, when both x_i and x_j are 1, $\Pr(x_i x_j = 1) = \Pr(x_i = 1, x_j = 1) = k/N * (k - 1)/(N - 1) = (k(k - 1))/(N(N - 1))$. As $E[X_i]$ and $E[X_j]$ are both k/N, we get the covariance as $(k(k - 1))/(N(N - 1)) - (k/N)^2$. Take $N^2(N - 1)$ as a common denominator to get $\text{Cov}(x_i, x_j) = -k(N - k)/[N^2(N - 1)]$.

The coefficient of skewness is as follows:

$$\beta_1 = \frac{(N - 2k)(N - 2n)\sqrt{N - 1}}{(N - 2)\sqrt{nk(N - k)(N - n)}}. \tag{7.20}$$

Inverse moments are useful in life-testing problems, evaluating risks of estimators and in Stein estimation [31]. The first inverse moment is

$$E(1/(X+1)) = \left(1/\binom{N}{n}\right) \sum_x (1/(x+1)) \binom{k}{x} \binom{N-k}{n-x}. \tag{7.21}$$

Multiply numerator and denominator by $(k+1)$, and take $1/(k+1)$ outside the summation to get

$$E(1/(X+1)) = \left(1/\left[(k+1)\binom{N}{n}\right]\right) \sum_x \binom{k+1}{x+1} \binom{N-k}{n-x}. \tag{7.22}$$

Note that the lower limit of this summation need not always be zero (when $N-k < n$). But when $n < N-k$, we could apply Vandermonde convolution (by supplying (adding and subtracting) the missing terms) and express it in terms of binomial coefficients. Higher-order inverse moments $E(1/(X+1)^r) > 1$ can be found using the relationship between rising factorials, signless Stirling numbers of the first kind $s(j,r)$ (Phillips and Zhigljavsky (2014) [63])

$$1/(X+1)^r = \sum_{j=r}^{\infty} s(j,r)/(X+2)^{[j]}. \tag{7.23}$$

7.4.3   GENERATING FUNCTIONS
The PGF can be expressed in terms of hypergeometric series as

$$P_x(t) = \binom{N-k}{n} / \binom{N}{n} \, _2F_1(-n,-k; N-k-n+1; t), \tag{7.24}$$

where $k = Np$. Take the constants to the denominator and express it also as hypergeometric function to get

$$P_x(t) = \, _2F_1(-n,-k; N-k-n+1; t)/_2F_1(-n,-k; N-k-n+1; 1). \tag{7.25}$$

Using a property of hypergeometric series this could also be written as $P_x(t) = \, _2F_1(-n,-k; -N; 1-t)$. The MGF does not have simple form, but is expressed in terms of hypergeometric functions as

$$M_x(t) = \binom{N-k}{n} / \binom{N}{n} \, _2F_1\left(-n,-k; N-k-n+1; e^t\right). \tag{7.26}$$

See Table 7.1 for more properties.

Theorem 7.1   *If X and Y are independent* BINO(m, p) *and* BINO(n, p) *random variables, then the distribution of $X|X+Y = n$ is hypergeometric, and is independent of p.*

Table 7.1: Properties of hypergeometric distribution

Property	Expression	Comments
Range of X	$x = \max(0, n - N + k),\ldots, \min(k, n)$	Discrete, finite
Mean	$\mu = nk/N$	$\sigma^2 < \mu$
Variance	$\sigma^2 = (nk/N)(1 - k/N)(N - n)(N - 1)$	$= \mu\,(1 - k/N)(N - n)(N - 1)$
Mode	$\lfloor (k + 1)(n + 1)/(N + 2) \rfloor$	
Skewness	$\gamma_1 = \dfrac{(N - 2k)(N - 2n)(N - 1)^{1/2}}{\lfloor nk(N - k)(N - n)^{1/2}\,(N - 2)}$	
CV	$\{(N - k)(N - n)/[nk(N - 1)]\}^{1/2}$	
Cov(x_i, x_j)	$\mathrm{Cov}\,(x_i, x_j) = -\,k(\frac{N - k}{[N^2(N - 1)]})$	
MD	$2x(N - k - n + x)\binom{k}{x}\binom{N - k}{n - x}/[N\binom{N}{n}]$	$2\,\mu_2 f_m[1 + 1/N]$
Factorial moments	$\mu_{(r)} = E[x_{(r)}] = (k)_r\binom{N - r}{n - r}/\binom{N}{n}, r = 1,2,\ldots, \min(k,n)$	$(k)_r\,(n)_r/(N)_r$
Recurrence	$f(x)/f(x-1) = (n-x+1)(k-x+1)/[x(N - n - k + x)]$	

Symmetric when $N/2 = k$ or n. Write $(N - n/(N - 1)$ as $1 - (n - 1)/(N - 1)$ to get another expression for variance. MGF is $\binom{N-k}{n}/\binom{N}{n}\,{}_2F_1(-n, -k; N - k - n + 1; e^t)$.

Proof. Consider the random variable $Z = X + Y$. As the probability p is the same, this is distributed as $\mathrm{BINO}(n + m, p)$. The conditional distribution of X given $Z = k$ is $P[X = x|Z = k] =$

$$P[X = x \cap Z = k]/P[Z = k] = P[X = x] * P[Y = k - x]/P[X + Y = k]. \qquad (7.27)$$

As X and Y are independent, we have

$$\binom{m}{x}p^x q^{m-x} \binom{n}{k - x}p^{k-x}q^{n-k+x} / \binom{m + n}{k}p^k q^{m+n-k}. \qquad (7.28)$$

This reduces to $\binom{m}{x}\binom{n}{k-x}/\binom{m+n}{k}$. This obviously is independent of p.   □

Problem 7.2   If X and Y are independent $\mathrm{NBINO}(m, p)$ and $\mathrm{NBINO}(n, p)$ random variables, then the distribution of $X|X + Y = n$ is negative hypergeometric, and is independent of p.

7.4.4   MEAN DEVIATION

The MD is most easily found using the MDGF (Chattamvelli and Shanmugam (2019) [18], pp. 43–45) as $D_x(t) =$

$$2E\left(t^x/(1 - t)^2\right) = 2\binom{N - k}{n} / \left[\binom{N}{n}(1 - t)^2\right]\,{}_2F_1(-n, -k; N - k - n + 1; t), \qquad (7.29)$$

as the coefficient of $t^{\lfloor \mu \rfloor - 1}$ where μ is the mean with a correction term added when the mean is not an integer. Alternately, it can be expressed in terms of $m = \lfloor \mu + 1 \rfloor$ as (Kamat (1965) [41])

$$\text{MD} = 2m(Nq - n + m)/N \binom{Np}{m}\binom{Nq}{n-m}/\binom{N}{n}. \tag{7.30}$$

7.5   TRUNCATED HYPERGEOMETRIC DISTRIBUTIONS

The zero-truncated hypergeometric (ZTH) distribution is used in many fields including CMR models. It is also used in validation studies that involve two independent processes. Consider a large election (say nationwide) in which the votes are counted electronically and manually. An election audit takes a sample of both counts and tally it to see if there are any mismatches in the results. Possible mismatches may result in larger audit or recounts. $\Pr[X = 0]$ in this case indicates that there were no bugs in the two methods.

The PMF is $f(x; p, N, n) =$

$$\binom{Np}{x}\binom{N-Np}{n-x}\Big/\left[1 - \binom{N-Np}{n}\Big/\binom{N}{n}\right]\binom{N}{n}$$

$$= \binom{Np}{x}\binom{Nq}{n-x}\Big/\left[\binom{N}{n} - \binom{Nq}{n}\right], \tag{7.31}$$

where $x = 1, 2, \ldots, \min(n, k)$.

Left-truncated hypergeometric distribution is obtained by truncating at a positive integer K. A right-truncated hypergeometric distribution is also possible, but of limited utility.

If truncation occurs in the left tail at $x = 0$, and the right tail at n, the PMF becomes

$$\binom{Np}{x}\binom{N-Np}{n-x}\Big/\left[1 - \left[\binom{Np}{n} + \binom{N-Np}{n}\right]\Big/\binom{N}{n}\right]\binom{N}{n}$$

$$= \binom{Np}{x}\binom{Nq}{n-x}\Big/\left[\binom{N}{n} - \binom{Np}{n} - \binom{Nq}{n}\right]. \tag{7.32}$$

7.5.1   APPROXIMATIONS FOR HYPERGEOMETRIC DISTRIBUTION

Hypergeometric probabilities can be approximated by the binomial distribution. When N and k are large, $p = k/N$ is not near 0 or 1, and n is small w.r.t. both k and $N - k$, the HGD is approximately a $\text{BINO}(n, k/N)$. Symbolically,

$$\lim_{N,k\to\infty, k/N\to p} \binom{k}{x}\binom{N-k}{n-x}\Big/\binom{N}{n} = \binom{n}{x}p^x(1-p)^{n-x}. \tag{7.33}$$

Four different binomial approximations exist due to the symmetries (i) $f(x;k,n,N) = f(x;n,k,N)$, (ii) $f(x;k,n,N) = f(k - x;k, N - n, N)$, and (iii) $f(x;k,n,N) = f(n - x; N - k, n, N)$, mentioned above. If k/N is small and n is large, the probability can be approximated using a Poisson distribution. Closed-form expressions for tail probabilities do not exist, except for particular values of the parameters. But, in general,

$$F_x(x;m,n,N) = F_x(y; N - m, N - n, N) \quad \text{where} \quad y = N - m - n + x. \quad (7.34)$$

7.6 APPLICATIONS

The hypergeometric distribution finds applications in a large number of fields. It is used in acceptance sampling of SQC, protein identification in bioinformatics, pathway enrichment in genomics, patient selection in medical sciences, etc. The capture-recapture model described below is used to estimate unknown population size and proportions [46]. Medical informatics uses CMR techniques for many purposes as described below. It is also used in gene selection studies and as invalidation test for binary forecasts [7].

7.6.1 HYPERGEOMETRIC TESTS

These are statistical tests that use the hypergeometric distribution to calculate the significance that a sample drawn from an unknown population indeed belongs to this distribution. Examples are median test, run test, and Fisher's exact test used in contingency tables. The Pearson's χ^2 test for 2×2 contingency tables is an approximate one, and not recommended when frequencies are small.

7.6.2 FISHER'S EXACT TEST

Fisher's exact test is used to analyze 2×2 contingency tables using row and column totals as parameters of hypergeometric distribution even for small frequencies. Samples are assumed to have been drawn from a hypergeometric distribution whose parameters are the row and column totals. Suppose $Y = Y_1, Y_2, \ldots, Y_n$ are a sequence of IID Bernoulli random variables with the same parameter p, and the values assumed by each of them are $\{0, 1\}$. When Y is dependent on explanatory variables $X_1, X_2, \ldots, X_n$, we could build a logistic model $\Pr(Y_j = 1) = \exp(a + bx_j)/[1 + \exp(a + bx_j)]$ where a and b are unknown constants. Consider the problem of testing $H_0 : b = 0$ against the one-sided alternative $H_1 : b > 0$. The significance level for testing this hypothesis is given by the tail area of the hypergeometric distribution. Similarly, tests for interaction between rows and columns uses the test statistic $t = p_{11}p_{22}/[p_{12}p_{21}]$ where p_{jk} for $j, k \in \{1, 2\}$ denotes the probability that an observation belongs to (j, k)th cell in a 2×2 contingency table. If the null hypothesis is $H_0 : t = 1$ (of no interaction between rows and column attributes) against the one-sided alternative $H_1 : t < 1$ (negative association) is again the sum of the probabilities of a hypergeometric distribution. Fisher's 2×2 contingency test was extended to $r x c$ contingency table by Carr (1980) [12] and Mehta and Patel (1983) [55].

7.6.3   MEDIAN TEST

As the name implies, this test is used to compare the observed data values with the population median. Typically, we apply this test on an experimental (treatment) group E and a control group C. The median of the control group is found first. The variable of interest is the number of observations in the treatment group that are less than M.

7.6.4   RUNS TEST

This test uses the occurrence of "success runs" of length k. A success run is a contiguous occurrence of several successes together. Let R denote the number of success runs of length k:

$$
\Pr(R = 2x) = 2\binom{k-1}{x-1}\binom{N-k-1}{x-1}/\binom{N}{k},
$$

$$
\Pr(R = 2x+1) = \left[\binom{k-1}{x}\binom{N-k-1}{x-1} + \binom{k-1}{x-1}\binom{N-k-1}{x}\right]/\binom{N}{k},
$$

(7.35)

where k is the number of successes out of N. Then the hypergeometric distribution can be used to construct the test (Guenther, 1978)[29].

7.6.5   ACCEPTANCE SAMPLING

Several manufacturing systems are moving toward automation due to the rising human labor. Moreover, automation can reduce the defect rate in high technology areas like semiconductor and nanotechnology industries. Inspecting all items in a manufactured lot may be either expensive or impossible due to the destructive nature of defect testing. In such cases the manufacturer allows customers to do acceptance testing, in which customers test each item in a received shipment for acceptance at their end and either accepts a lot or rejects it. If the customer keeps track of the number of defectives and conveys it to the manufacturer, they can come up with a frequency distribution of the defects in past shipments. If p denotes the average of the past defects, it can be used to model the number of defectives in a sample of size n taken from N newly manufactured items as

$$
f(x; n, p, N) = \binom{Np}{x}\binom{N-Np}{n-x}/\binom{N}{n}.
$$

(7.36)

If the probability distribution of defectives is denotes as $\{\pi_0, \pi_1, \ldots, \pi_N\}$ where π_k denotes the number of shipments with k defectives, the joint distribution of defects given the defect rate is

p is given by

$$f(x; n, p = \pi, N) = \pi \binom{Np}{x} \binom{N - Np}{n - x} / \binom{N}{n}. \tag{7.37}$$

Problem 7.3   A rocket booster uses a component part supplied by an external vendor. The vendor has found from in-house testing that 3 in 50 parts may fail due to extreme temperature variations. The booster assembly needs 8 booster parts that are selected from 12 parts supplied by the vendor. The booster will work satisfactorily if 6 or more of them works as expected.

7.6.6   THE CMR MODEL

The CMR model, introduced in Chapter 5, is a popular technique used in ecology, epidemiology, marine biology, mineral geology, microscopy, and many other fields. It is increasingly being used in novel areas like opinion polls, marketing research, intrusion detection (in computer and wireless networks, IoT connected gadgets), informatics, national security, to name a few. A fundamental question that concerns scientists is "How best to estimate the population size of entities in a field?." The CMR model can be used to estimate an unknown population size using simple random sampling twice as follows: (i) capture a small random sample of m items. (ii) mark (or tag) the captured items. (iii) release the m items into the population. (iv) allow it to be well mixed with the population, and (v) recapture a slightly larger sample of size n. The items mentioned above are quite often wild animals, birds, or insects (like butterflies, bees) in ecology; fish, turtle, and other moving marine creatures in biology; humans (usually adults) in opinion polls, marketing research; viruses and other malware in intrusion detection; piece of information in the form of research articles, alternate knowledge like folklore medicines in informatics; pathogens in microscopy, unusual events, or occurrences (like social media posts) in national security. It may also be contaminants in soil samples (usually in motion) in soil chemistry, mineral deposits in geology and petroleum engineering. It is implicitly assumed in this model that the location of each entity is the realization of a stochastic process. In addition, it is assumed that the: (i) population size is finite and is more or less steady; (ii) marked and released items are not dead (at least until the second sampling is done) and new offspring are negligible; (iii) marked items are healthy enough to well-mix randomly with the population (in the case of animate entities);[3] (iv) the time-gap between the first and second sampling is reasonable for some marked items to reach the population horizons, and not long enough for the population size to change appreciably; (v) the location of sampling and resampling provides an equal chance for every item in the population to be included; and (vi) there is no violation of closure assumptions. In other words, the sampling need to be decided based upon the extend of the area where the population resides, and the sampling time interval calculated to ensure some

---

[3]The CMR model has also been applied to estimate population abundance of inanimate entities like contaminants, mineral deposits, etc.

degree of randomness. If it is a pond or lake where fish is the entity, sampling and releasing can happen at the center or at various points on the shore. But if the habitat is spread over thousands of kilometers (as in tiger and other wildlife conservations in wildlife ecology), the sampling and releasing can take place at multiple locations spread uniformly within the habitat or other types of sampling (stratified or cluster sampling when obvious habitat density differences exist or it is dynamic) employed.

The violation of closure assumption occurs when the habitat spreads or shrinks in-between the first and second sampling. This often happens when the subjects are humans, and rarely due to natural reasons like forest fires, global warming, polar ice melts, season changes, etc. Changes in population abundance occurring over space, time, or both due to ecological and other factors can be measured using multiple runs of the CMR models. Advances in technology that use radio-tagging, GPS positioning, aerial surveillance using drones, remote video cameras, and smart-dust allow locating inactive (stationary) tagged items, understand dynamic behavior of habitat density differences, and helps to improve the population abundance estimate.

Without loss of generality, it is assumed that the size of the second sample (n) is greater than or equal to m (number of marked entities). Let $x_j = 1$ if jth item in the second sample is marked and $x_j = 0$ otherwise. Then count the number of marked items k in our sample of size n and equate it to the population mean to get $k = m * n/N$ from which the point estimate as $\hat{N} = mn/k$, which is called the abundance estimate of the population. Literally, this can be stated as follows:

```
Population size = number of marked items times number of sampled
  items divided by the number of marked items in the sampled items.
```

If it is not an integer, the next higher integer (called ceil as $N = \lceil m * n/k \rceil$) is taken as the population size. Improvements to this estimate has been suggested by Chapman (1951) [13] as $\hat{N} = [\lfloor (m + 1)(n + 1)/(k + 1) \rfloor] - 1$, Robson and Regier (1964) [68], and Seber (2002) [73] as $\hat{N} = [\lfloor (m + 2)(n + 2)/(k + 2) \rfloor]$. As $(m + 1)(n + 1)/(k + 1) = (mn/k)[k/(k + 1)] + (m + n + 1)/(k + 1)$, the improved estimates are higher for large k.

Other models (like regression models) can be applied to refine the estimate on an ongoing basis if the population size is not steady, but grows or shrinks over time.

The standard model assumes homogeneity among the distribution of individual entities. This allows us to model the head-count of entities in a sub-region as a random variable. This may not always hold due to drought, flood, forest fires, and other natural calamities on land, oil-spills, toxic chemicals, or radioactive dumps in water; or herding of groups of entities due to food abundance, or predator attacks. It is a tacit assumption that the second sample (of size n) has enough of the marked items. No reasonable conclusions can be made (except that the population size is very large) in the unusual case when this sample does not have any marked items present (on the other extreme, if all marked items m are present in the second sample, it is an indication that the population size is close to the sample size n, or the marked items failed to mix randomly). A solution is to continue sampling until at least one of the marked items are present. In this case, the samples size (X) has a geometric distribution. Even the presence of just one marked item in a sample of size n can result in biased estimates. Ideally, we would like to continue sampling until a fixed number k of marked items are present. In this case the number of samples taken has a negative hypergeometric distribution (Chapter 8) because the probability of obtaining $k - 1$ marked items in $x + k - 1$ trials followed by a marked item in $(x + k)$th trial is $f(x, k, n, N) =$

$$\binom{x + k - 1}{k - 1} (m(m - 1) \dots (m - k + 1)/[N(N - 1) \dots (N - k + 1)]) [(N - m)/N]^x. \quad (7.38)$$

7.6.7 LINE-TRANSECT MODEL

The line-transect model (LTM) is a rival that assumes heterogeneity of sampled data. LTM and point-transect model (PTM) are distance-based techniques to estimate the density or abundance of unknown populations. In this method, replicate transect lines are randomly located in the space of the habitat. Data on distances of objects being sampled are then collected. As the observer moves along a random straight line, polar coordinate based methods are often used to record the distance of an object from the current observer position. If the motion of the object is slow or it is stationary, this can be implemented using drones, multiple high-resolution cameras or a pool of smart-dust. All objects that lie on the transect line will have $\theta = 0$, and will be detected. Otherwise, θ is the angle between the transect line and an imaginary line connecting the observer and the object, and r is the radial distance. The probability of detection decreases with increasing distance from the transect line. Average density of objects within the region are then calculated as $n/(P * A)$, where n is the count of objects detected, P is the probability of detection of an object within a width w of the line, and A is the area of the feasible region covered. The tailing-off of the detectability is modeled using detection functions (half-normal, half-T, negative exponential, or some other continuous function).

7.6.8 HIGHWAY ENGINEERING

Highway information systems are used to get up-to-the-minute information on vehicular movements, congestion spots, accident spots, and related information. Past data on these are valuable in such models. Highway engineers identify accident hotspots that are locations where many accidents took place. This induces a dichotomy among hundreds of points (locations) of interest. Traffic flows are examined on an ongoing basis and a small sample of current troublesome points (where traffic congestion is just building up or is beyond a threshold) are identified. The number of hotspots in the sample has a binomial distribution if the points are selected randomly, and a hypergeometric distribution if it is done one-by-one without replacement.

The same concept can be used in similar networks like computer networks, intranets, and the like.

7.6.9   MEDICAL SCIENCES

The hypergeometric distribution has multiple applications in medical sciences. Some epidemiological studies take data from several sources and eliminate duplicates to come up with research data. Such data are likely to be biased downwards because an unknown portion of the population may be totally misrepresented. It is a known fact that some people are sensitive to particular medicines (like penicillin), or medical procedures. Suppose a hospital wishes to estimate the number of people in a city who are sensitive to such a medicine or procedure. Using past hospital data of patient visits, they can estimate the approximate percent of people of interest (for each such medicine or procedure). Assume that the funding and manpower available does not allow them to do a complete survey of the populace, but can contact at most n people only. Let X denote the number of people who are sensitive in a sample of size n (without replacement). Then X has a hypergeometric distribution.

The recent outbreak of the COVID-19 pandemic pointed out flaws in disease monitoring and imperfect surveillance systems. The CMR model is most appropriate when the population is boundary constrained. This means that multiple CMR models can be used for various cities and regions. As a large portion of the population in big cities are dynamic (they move out and move in) macro models may have to be built by identifying larger boundaries. A single-source CMR model may be biased in estimating the true number of people affected.

As another example, consider a stopping rule to be used to estimate the closeness to capturing the total body of literature on a given topic. This may be needed when searching for potentially relevant articles to be included in systematic reviews preparation using large medical databases like Medline, EMBASE, CINAHI, EBM reviews, etc. [Kastner et al., 2009] [42]. The CMR rule can be used as a stopping strategy to estimate the total number of articles in the subdomains involved. This may be done at multiple levels like title and abstract level, keywords, citation links, and full text levels. Information retrieval techniques like latent semantic indexing (LSI) may be used for most relevant document or article searches using the above levels [Chattamvelli (2016) [16]]. Number of articles to be further searched in the above-mentioned databases substantially comes down when multiple levels are combined.

7.6.10   BIOINFORMATICS

Proteins play an important role in cellular, metabolic, structural, and physiological mechanisms of living organisms. They are represented as binary vectors in which the ith bit position is a "1" if the ith genome contains a homolog of the corresponding gene, and is 0 otherwise. The length of a protein vector is the number of available genomes in it. This is called the phylogenetic profile in bioinformatics, which is a useful tool in personalized medicine, genomic medicine, and functional genomics studies. Major topics of interest in bioinformatics include protein identification, understanding the function of proteins, predicting the functions of uncharacterized proteins from their sequence, and getting insights into a protein's metabolism [71]. All of these are done using protein databases that contain a large number of already sequenced proteins, some of which are still un-annotated. Let X denote the number of genomes containing a homolog of the corresponding gene in a protein of length n. Then X has a hypergeometric distribution.

7.7 SUMMARY

This chapter introduced the hypergeometric distribution, its properties and various applications. We have discussed only the classical hypergeometric distribution, although there exist many varieties of them. Interested reader is referred to Johnson et al., (2005) [39], Balakrishnan and Nevzorov (2003) [5], and other references in the bibliography.

CHAPTER 8

Negative Hypergeometric Distribution

After finishing the chapter, readers will be able to . . .

- Explain negative hypergeometric distributions.

- Explore the properties of negative hypergeometric distribution.

- Describe beta-binomial distribution and its properties.

- Apply negative hypergeometric distribution in practical situations.

8.1 DERIVATION

The classical hypergeometric distribution (HGD) is the distribution of x successes in n trials without replacement from N items containing k successes, and $N - k$ failures. Note that the number of trials n is fixed. It was mentioned in the CMR model of the application section of last chapter that we may have to continue sampling one at a time without replacement from a finite population until we get r (fixed) marked items. Practical experiments that result in a negative hypergeometric distribution can be characterized by the following properties:

1. The experiment consists of sampling *without replacement* from a dichotomous population.

2. Trials can be repeated independently n times under identical conditions.

3. The probability of occurrence of the outcomes p_k varies (increases) from trial to trial until the experiment is over.

4. The random variable X denotes the number of times one of the dichotomous groups is selected until the other count becomes r.

As an example, consider the problem of drawing cards without replacement from a well-shuffled deck of cards until r cards of a fixed suite (say, five spades) are selected. Bulk manufacturing companies use it for acceptance sampling where success denotes that a manufactured product is defective. Let X denote the number of trials needed to get r successes when sampling without replacement from a population of size N containing k successes. Then X has a negative (or inverse) hypergeometric distribution (NHGD)[1] with parameters k, r, N denoted as NHG(k, r, N). As mentioned in Chapter 6, successes and failures may denote "marked" and "unmarked" in ecology and biological sciences, "working" and "non-working" (defective) in engineering sciences, "conforming" and "non-conforming" in SQC, etc. Whereas the negative binomial distribution (NBINO) models waiting times with replacement, the NHGD models waiting times without replacement. The difference between positive HGD and NHGD is that the number of trials is fixed in HGD, whereas the number of successes is fixed (and the number of trials is varying) in NHGD. Similarly, the range is infinite for NBINO, while it is finite for NHGD.

As done in NBINO, let x denote the number of failures to get r successes. Then there exist $r - 1$ successes in $x + r - 1$ trials, and $(x + r)$th trial is a success. The probability of the first part can be found using positive HGD (because sampling is done without replacement). Probability of the second part is the ratio of the number of successes remaining in the population divided by the remaining population size. As these two events are independent, the desired probability is given by their product as

$$f(x; k, r, N) = \left(\binom{k}{r-1} \binom{N-k}{x} \middle/ \binom{N}{x+r-1} \right) \times \frac{k - (r-1)}{N - (x + r - 1)}. \tag{8.1}$$

Expand the RHS, and rearrange as

$$\frac{k!(k - r + 1)}{(r - 1)!(k - r + 1)!} \binom{N-k}{x} (x + r - 1)! \frac{(N - x - r + 1)!}{N!(N - x - r + 1)}. \tag{8.2}$$

Cancel out $(k - r + 1)$ in the numerator, and $(N - x - r + 1)$ in the denominator to get

$$f(x; k, r, N) = (k/N) \binom{k-1}{r-1} \binom{N-k}{x} \middle/ \binom{N-1}{x+r-1}. \tag{8.3}$$

8.1.1   DUAL NHGD

Swapping the successes and failures results in the dual NHGD (DNHGD). Now there exist $r - 1$ failures in $x + r - 1$ trials, and $(x + r)$th trial is a failure. The probability of the first part can be found using positive hypergeometric distribution (because sampling is done without replacement). Probability of the second part is the ratio of the number of failures remaining in the population divided by the remaining population size. As these two events are independent, the desired probability is given by their product as

$$f(x; k, r, N) = \left(\binom{k}{x} \binom{N-k}{r-1} \middle/ \binom{N}{x+r-1} \right) \times \frac{N - k - (r-1)}{N - (x + r - 1)}. \tag{8.4}$$

---

[1]also called hypergeometric waiting-time, beta-binomial, or Markov–Polya distribution.

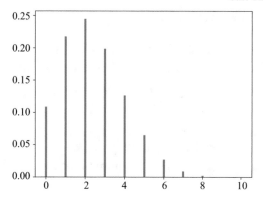

Figure 8.1: PMF of NHGD(5, 20, 30).

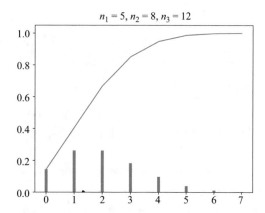

Figure 8.2: PMF of BBD ($n_1 = 5, n_2 = 8, n_3 = 12$).

As done above, this simplifies to

$$f(x; k, r, N) = (r/N) \binom{k}{x} \binom{N-k}{r} / \binom{N-1}{x+r-1}. \tag{8.5}$$

8.2   ALTERNATE FORMS

As in NBINO, we could consider x as the total number of outcomes or trials (successes and failures) so that in $x - 1$ trials there are $r - 1$ successes and $x - r$ failures, and the xth trial results in another success. This gives the PMF as

$$f(x;k,r,N) = \left(\binom{k}{r-1}\binom{N-k}{x-r}/\binom{N}{x-1}\right) \times \frac{k-(r-1)}{N-(x-1)}$$

$$= \frac{k}{N}\binom{k-1}{r-1}\binom{N-k}{x-r}/\binom{N-1}{x-1},$$

$$(8.6)$$

where the last expression is obtained after canceling out $k - r + 1$ and $N - x + 1$. Expand the RHS as factorials, and use $k(k-1)! = k!$ (in the numerator), and $N(N-1)! = N!$ (in the denominator) to get

$$\frac{k!}{(r-1)!(k-r)!} \frac{(N-k)!}{(x-r)!(N-k-x+r)!} \frac{(x-1)!(N-x)!}{N!}.$$

$$(8.7)$$

Rearrange the terms as

$$\frac{(x-1)!}{(r-1)!(x-r)!} \frac{(N-x)!}{(k-r)!(N-x-k+r)!} \frac{k!(N-k)!}{N!}.$$

$$(8.8)$$

This using combination symbols is

$$f(x;k,r,N) = \binom{x-1}{r-1}\binom{N-x}{k-r}/\binom{N}{k}.$$

$$(8.9)$$

For the first combination term to be valid, x must be $\geq r$, and for the second to be valid, $N - x$ must be $\geq k - r$. From this we get $r \leq x \leq N - k + r$. ·

Other forms that use $k = Np$ and $N - k = Nq$ can be obtained from above representations, one of which from (8.5) is as follows:

$$f(x;r,p,N) = (r/N)\binom{Np}{x}\binom{Nq}{r}/\binom{N-1}{x+r-1}.$$

$$(8.10)$$

If r denotes the number of failures, and x denotes the number of successes when the experiment is stopped, the PMF could alternatively be obtained using the following argument. In $(x + r - 1)$ trials there are $r - 1$ failures and x successes. This can be found using hypergeometric distribution as $\binom{k}{x}\binom{N-k}{r-1}/\binom{N}{x+r-1}$. Total number of failures remaining is $(N-k) - (r-1) = N - k - r + 1$, and the size of the remaining population is $N - (x + r - 1)$. As these two events are independent, the probability is the product $\left(\binom{k}{x}\binom{N-k}{r-1}/\binom{N}{x+r-1}\right) \times (N - k - r + 1)/(N - x - r + 1)$, as in (8.4). Expand the RHS as factorials

$$\frac{k!}{x!(k-x)!} \frac{(N-k)!}{(r-1)!(N-k-r+1)!} \frac{(x+r-1)!(N-x-r+1)!}{N!} \frac{N-k-r+1}{N-x-r+1}.$$

$$(8.11)$$

Cancel out $(N - k - r + 1)$ and $(N - x - r + 1)$, and rearrange the terms as

$$\frac{(x + r - 1)!}{(r - 1)!x!} \frac{(N - x - r)!}{(k - r)!(N - x - k)!} \frac{k!(N - k)!}{N!}, \tag{8.12}$$

which simplifies to

$$f(x; k, r, N) = \binom{x + r - 1}{x}\binom{N - r - x}{k - r} / \binom{N}{k}$$

$$= \binom{x + r - 1}{r - 1}\binom{N - r - x}{k - r} / \binom{N}{k}, \tag{8.13}$$

where $x = \{0, 1, 2, \ldots, k\}$ is the total number of successes. Put $y = x + r$ to get a compact representation as

$$f(y; k, r, N) = \binom{y - 1}{y - r}\binom{N - y}{k - r} / \binom{N}{k} \quad \text{for} \quad y = r, r + 1, \ldots, r + k. \tag{8.14}$$

A dual of this could also be obtained by swapping the roles of successes and failures. Let $k = Np$ denote the number of successes so that $Nq = N - k$ is the number of failures. In $x - 1$ trials there are $r - 1$ successes, and $x - r$ failures with PMF $f(x; p, r, N) =$

$$\left(\binom{Np}{r - 1}\binom{Nq}{x - r} / \binom{N}{x - 1}\right) \times \frac{Np - (r - 1)}{N - (x - 1)}. \tag{8.15}$$

This simplifies to

$$\binom{x - 1}{x - r}\binom{N - x}{Nq - x + r} / \binom{N}{Nq}. \tag{8.16}$$

Beta-binomial distribution (BBD) discussed below is a special case when x and $(n - x)$ are integers. Another version of (8.13) in terms of n_1, n_2, n_3 has PMF

$$\binom{n_1 + x - 1}{x}\binom{n_2 + n_3 - n_1 - x}{n_3 - n_1} / \binom{N}{n_2} = \binom{n_1 + x - 1}{n_1 - 1}\binom{N - n_1 - x}{n_1 - x} / \binom{N}{n_2}, \tag{8.17}$$

where $N = n_2 + n_3$, for $x = 0, 1, 2, \ldots, n_2$. This reduces to the DUNI($\binom{N}{n_2}$) for $n_1 = n_3 = 1$. This has mean $\mu = n_1 n_2 / (n_3 + 1)$.

This distribution can be obtained from Polya's urn model as follows. Consider an urn containing two types of balls (say r red and w white balls). Balls are drawn one-by-one at random without looking at the color. Assume that after noting down the color of the selected ball, we return c *additional* number of balls of the same color to the urn, where c is an integer (positive

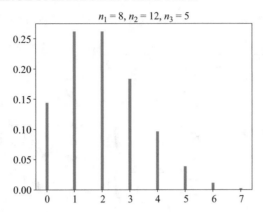

Figure 8.3: PMF of BBD(8, 12, 5).

or negative). As mentioned in last chapter, different distributions arise for different values of c. *No balls* are returned back including the selected ball for $c = -1$. Removing balls until a specified number r balls of a specific color is selected results in either negative binomial (with replacement) or negative hypergeometric (without replacement) distributions. The case $c = 1$ in which one *additional* ball of the same color is returned to the urn also results in a NHGD (Irwin (1954) [34], Kemp and Kemp (1956) [43]).

8.2.1   RELATION TO OTHER DISTRIBUTIONS

This distribution has the same type of relationship with NBINO, as HGD is related to BINO(n, p) (see Table 8.1). If the number of items selected (n) is one, we have a Bernoulli distribution with $p = k/N$. This distribution can be approximated by a binomial distribution, Poisson distribution [102], and negative binomial distribution ([64], [33], [103]). For large values of N and k, the NHGD tends to NBINO(r, p) where $p = k/N$. Similarly, when number of failures $N - k \to \infty$, DNHGD can be approximated by NBINO(r, p). Instead of waiting for one type of event (say successes) to occur r times, we could wait for type-1 event (or object) to occur r times *and* type-2 event to occur s times to get a "later waiting time" distribution; and type-1 event to occur r times *or* type-2 event to occur s times to get a "sooner waiting time" distribution [51], [40]. The Dirichlet multinomial distribution is a multivariate analogue of NHGD.

It is related to the binomial error model as well. In psychological tests, if the true score is approximately linearly related (through binomial error regression) with observed scores, then observed score is NHGD [53].

Table 8.1: Relationships among some univariate distributions

X Denotes	With Replacement	Without Replacement
# of successes	Binomial	Hypergeometric
# of trials for r successes	Negative Binomial	Negative Hypergeometric
Among these, only negative binomial distribution has infinite range.		

8.3 PROPERTIES OF NHGD

This distribution has three parameters, all of which are integers (in the classical distribution). The PMF could become numerically intractable for some parameter values due to large numbers that may appear in the combination terms. A solution is to represent the PMF in terms of Pochhammer notation, cancel out common factors from numerator and denominator, and compute the terms iteratively (instead of computing the numerator and denominator separately) (see [28], [29], [44]). A recurrence for (8.9) can be obtained as $f(x; k+1, r, N) = ((k+1)(N+r-x-k)/[(N-k)(k+1-r)])\, f(x; k, r, N)$, from which the mode can be found as $\lfloor r(N+1)/k \rfloor$. Similar recurrences can easily be obtained for other representations.

Good approximations are available for certain parameter ranges. Poisson approximations may be used when N is large and $r/(N-k)$ is small. The accuracy increases when r approaches k.

The ratio $\mathrm{IJR} = \Pr[X = 1]/\Pr[X = 0]$ for the negative hypergeometric distribution is $r(k-r)/(N-r)$. For displaced distributions this becomes $\mathrm{IJR} = \Pr[X = ll+1]/\Pr[X = ll]$ where ll is the lower limit.

Problem 8.1 A city has 20,000 COVID-19 recovered persons. A researcher wishes to recruit $r = 200$ of them for a cohort study. If the chance that a randomly chosen person had COVID-19 is $p = 5\%$, what is the distribution of the number x of non-infected persons contacted in random sampling without replacement until 200 recovered persons are recruited? If the chance that a randomly selected person agrees to participate is 15%, what is the probability that 300 persons must be asked before 200 are found to agree?

8.3.1 GENERATING FUNCTIONS

The PGF can be expressed in terms of hypergeometric function as

$$P_x(t) = t^r k_{(r)}/N_{(r)\,2}F_1(r, -(N-k); -(N-r); t), \tag{8.18}$$

which is a terminating series for r positive [44]. Take the constants to the denominator and express it also as hypergeometric function to get

$$P_x(t) = {}_2F_1(r, -(N-k); -(N-r); t)/{}_2F_1(r, -(N-k); -(N-r); 1). \tag{8.19}$$

The MGF does not have simple form, but is expressed in terms of hypergeometric functions as

$$M_x(t) = \left(\exp(tr)k_{(r)}/N_{(r)}\right) \, _2F_1(r, -(N-k); -(N-r); e^t). \tag{8.20}$$

8.3.2   MEAN AND VARIANCE

We will use (8.9) to derive the mean and variance. As the total probability should add up to one, we have $\sum_{x=r}^{N-k+r} \binom{x-1}{r-1}\binom{N-x}{k-r}/\binom{N}{k} = 1$. Cross-multiply and change the variable as $y = x - r$ to get

$$\sum_{y=0}^{N-k} \binom{y+r-1}{r-1}\binom{N-y-r}{k-r} = \binom{N}{k}. \tag{8.21}$$

Put $a = N - r, b = k - r, c = r - 1$ so that $a + c = N - 1, b + c = k - 1, a - b = N - k$ to get

$$\sum_{y=0}^{a-b} \binom{y+c}{c}\binom{a-y}{b} = \sum_{y=0}^{a-b} \binom{y+c}{y}\binom{a-y}{b} = \binom{a+c+1}{b+c+1}. \tag{8.22}$$

Write $x\binom{x-1}{r-1} = x(x-1)!/[(r-1)!(x-r)!] = rx!/[r!(x-r)!]$ (by multiplying numerator and denominator by r) $= r\binom{x}{r}$. Now $\mu = E(X) =$

$$\sum_{x=r}^{N-k+r} x\binom{x-1}{r-1}\binom{N-x}{k-r}/\binom{N}{k} = \sum_{x=r}^{N-k+r} r\binom{x}{r}\binom{N-x}{k-r}/\binom{N}{k}. \tag{8.23}$$

Change the variable as $y = x - r$, and then use the above identity to get the mean as

$$\mu = r\sum_{y=0}^{N-k} \binom{y+r}{r}\binom{N-r-y}{k-r} = r\binom{N+1}{k+1}/\binom{N}{k} = r\frac{N+1}{k+1}. \tag{8.24}$$

In comparison, the negative binomial mean is rN/k. As $r[N/k - (N+1)/(k+1)] = r(N - k)/[k(k+1)]$, the means converge when $k \to N$ (or $r \to 0$). Otherwise, the waiting time is smaller in sampling without replacement, as is to be expected. The expected value around r is $E(X - r) = r(N+1)/(k+1) - r = r(N-k)/(k+1)$, where $(N-k)$ is the number of failures in the population. The mean of the DNHGD in (8.13) is $rk/(N-k+1)$. To find the variance, write $x(x+1)\binom{x-1}{r-1} = (x+1)x(x-1)!/[(r-1)!(x-r)!] = r(r+1)(x+1)!/[(r+1)!(x-r)!]$ (by multiplying numerator and denominator by $r(r+1)) = r(r+1)\binom{x+1}{r+1}$. We will first derive the rising factorial moment $E[X(X+1)]$, and use $\sigma^2 = E[X(X+1)] - E(X) - [E(X)]^2 = E[X(X+1)] - E(X)[1 + E(X)]$. Multiply by $\binom{N}{k}$ to get $\binom{N}{k}E[X(X+1)] =$

$$\sum_{x=r}^{N-k+r} x(x+1)\binom{x-1}{r-1}\binom{N-x}{k-r} = \sum_{x=r}^{N-k+r} r(r+1)\binom{x+1}{r+1}\binom{N-x}{k-r}. \tag{8.25}$$

Take $r(r+1)$ outside the summation, and change the variable as $y = x - r$ to get

$$E[X(X+1)] = r(r+1) \sum_{y=0}^{N-k} \binom{y+r+1}{r+1}\binom{N-y-r}{k-r} / \binom{N}{k}. \tag{8.26}$$

Using the identity (8.22) this becomes $r(r+1)\binom{N+2}{k+2}/\binom{N}{k}$. Expand the combination terms and cancel out common factors to get $E[X(X+1)] =$

$$r(r+1)(N+1)(N+2)/[(k+1)(k+2)] = (r+1)_{(2)}(N+2)_{(2)}/(k+2)_{(2)}. \tag{8.27}$$

Then $\sigma^2 = E[X(X+1)] - E(X)[1 + E(X)] = r(r+1)(N+1)(N+2)/[(k+1)(k+2)] - r\frac{N+1}{k+1}[1 + r\frac{N+1}{k+1}]$. This simplifies to

$$\sigma^2 = r(N+1)(N-k)(k+1-r)/[(k+1)^2(k+2)]. \tag{8.28}$$

Put $r(N+1)/(k+1) = \mu$ to get $\sigma^2 = \mu(N-k)(k+1-r)/[(k+1)(k+2)] = \mu\frac{N-k}{k+2}(1 - \frac{r}{k+1})$. The CV of NHGD is $\sigma/\mu = \{[(N-k)(k+1-r)]/[r(N+1)(k+2)]\}^{1/2}$. The variance of dual NHGD in (8.13) can similarly be found as $\mu\frac{N+1}{N-k+2}(1 - \frac{r}{N-k+1})$, where $\mu = rk/(N-k+1)$ is its mean. Although the mean is linear in r, the variance is quadratic in r. A direct consequence of the linearity of the mean in r is that we could decompose the NHGD into r disjoint random variables $X_1, X_2, \ldots, X_r$, where X_1 denotes the number of trials needed to get the first success, X_2 denotes the *subsequent* number of trials needed to get the next success, and so on as done by Schuster and Sype (1987) [72]. A straightforward extension of their result is that the sum formed by any partition of the integer r will correspond to waiting time distributions. For instance, if $r = 3$, we could define "two-partition" as $X_1 + X_2$ where X_1 denotes the number of trials to get the first success, and X_2 is the subsequent trials to get the next two successes or *vice versa*. A "three-partition" as $X_1 + X_2 + X_3$ denotes three single-success waiting times (one after another). This shows that the NHGD is decomposable in $\phi(r)$ ways into sums of waiting-time random variables, where $\phi()$ denotes Euler's partition number.

Differentiate $r(k+1-r)$ as a function of r, and equate to zero to find the maximum as $(k+1) - 2r = 0$, or equivalently $r = (k+1)/2$. This shows that the variance is highest near $r = \lfloor(k+1)/2\rfloor$. Because of the symmetry of $r(k+1-r)$, the variance remains the same for r and $(k+1-r)$ successes. Although r is an integer, this distribution can be extended to non-integer case by expanding the combination terms, and replacing factorials by gamma function. The variance tends to zero when $r \to 0$ or $r \to (k+1)$. Variance of NHGD is less in comparison with the variance of NBINO.

8.3.3   FACTORIAL MOMENTS

Rising factorial moments are easier to find due to the $(x - 1)!$ in the numerator. The mth rising factorial moment is

$$\mu_{(m)} = E[x^{(m)}] = E[x(x + 1) \dots (x + m - 1)]$$

$$= \sum_{x=r}^{N-k+r} x(x + 1) \dots (x + m - 1) \binom{x - 1}{r - 1} \binom{N - x}{k - r} / \binom{N}{k}. \qquad (8.29)$$

Write this as $\binom{N}{k}\mu_{(m)} = $ •

$$\sum_{x=r}^{N-k+r} (x + m - 1)(x + m - 2) \dots (x + 1)x(x - 1)!/[(r - 1)!(x - r)!] \binom{N - x}{k - r}. \qquad (8.30)$$

Multiply numerator and denominator by $[(r + m - 1)(r + m - 2) \dots (r + 1)r]$, and write it as $(r + m - 1)_{(m)}$ to get $\binom{N}{k}\mu_{(m)} = (r + m - 1)_{(m)} \times$

$$\sum_{x=r}^{N-k+r} (x + m - 1)(x + m - 2) \dots (x + 1)x(x - 1)!/[(r + m - 1)!(x - r)!] \binom{N - x}{k - r}. \qquad (8.31)$$

Replace factorials by combination symbol to get

$$\binom{N}{k}\mu_{(m)} = (r + m - 1)_{(m)} \sum_{x=r}^{N-k+r} \binom{x + m - 1}{r + m - 1} \binom{N - x}{k - r}. \qquad (8.32)$$

Put $y = x - r$ to get

$$\binom{N}{k}\mu_{(m)} = (r + m - 1)_{(m)} \sum_{y=0}^{N-k} \binom{y + r + m - 1}{r + m - 1} \binom{N - y - r}{k - r}. \qquad (8.33)$$

Now use the identity in (8.22) to get $\binom{N}{k}\mu_{(m)} = (r + m - 1)_{(m)} \binom{N+m}{k+m}$ from which

$$\mu_{(m)} = (r + m - 1)_{(m)} \binom{N + m}{k + m} / \binom{N}{k} = (r + m - 1)_{(m)} \binom{N + m}{N} / \binom{k + m}{k}. \qquad (8.34)$$

This can be expressed in terms of falling Pochhammer notation as $\mu_{(m)} = (r + m - 1)_{(m)}(N + m)(N + m - 1) \dots (N + 1)/(k + m)(k + m - 1) \dots (k + 1)$. Put $m = 1$ to get $\mu_{(1)} = \mu = r(N + 1)/(k + 1)$, which is in complete agreement with (8.24). Next, put $m = 2$ to get $\mu_{(2)} = (r + 1)_{(2)}(N + 2)_{(2)}/(k + 2)_{(2)}$, which agrees with (8.27).

8.3.4   BETA-BINOMIAL DISTRIBUTION

This distribution is also called compound binomial or Markov–Polya distribution. It can be obtained from the binomial distribution in which the probability of success follows the beta law.[2] Consider the binomial random variable with PMF $f(x; n, p) = \binom{n}{x} p^x (1-p)^{(n-x)}$ where p is distributed as a type-I beta variable as $g_p(a, b) = (1/B(a, b)) p^{a-1} (1-p)^{b-1}$ for $0 \leq p \leq 1$. Since p is a continuous random variable in the range $(0, 1)$, we obtain the unconditional distribution of X by integrating out p from the joint probability distribution, and using the expansion $B(a, b) = \frac{\Gamma(a)\Gamma(b)}{\Gamma(a+b)}$ as

$$f_x(n, a, b) = \int_0^1 f(x|p) g(p) dp = \left(\binom{n}{x} / B(a, b) \right) \int_{p=0}^1 p^{x+a-1} (1-p)^{b+n-x-1} \, dp$$

$$= \binom{n}{x} \frac{\Gamma(a+b)}{\Gamma(a)\Gamma(b)} \frac{\Gamma(a+x)\Gamma(b+n-x)}{\Gamma(a+b+n)}. \tag{8.35}$$

By writing $\binom{n}{x} = n!/[x!(n-x)!] = \Gamma(n+1)/[\Gamma(x+1)\Gamma(n-x+1)]$, this can also be written as $f_x(n, a, b) =$

$$\Gamma(n+1)\Gamma(a+b)/[\Gamma(x+1)\Gamma(n-x+1)\Gamma(a)\Gamma(b)][\Gamma(a+x)\Gamma(b+n-x)/\Gamma(a+b+n)]. \tag{8.36}$$

This form is widely used in Bayesian analysis. This distribution reduces to the NHGD as shown below. Put $a = n_1, b = n_3 - n_1, n = n_2$, so that $a + b = n_3$, and rearrange it as follows:

$$\Gamma(n_1 + x)\Gamma(n_2 - x + n_3 - n_1)\Gamma(n_3)$$
$$\Gamma(n_2 + 1)/[\Gamma(n_2 + n_3)\Gamma(n_1)\Gamma(n_3 - n_1)\Gamma(x+1)\Gamma(n_2 - x + 1)]. \tag{8.37}$$

Now replace each gamma term by factorial using $\Gamma(n) = (n-1)!$, and rearrange to get

$$\frac{(n_1 + x - 1)!}{x!(n_1 - 1)!} \frac{(n_2 - x + n_3 - n_1 - 1)!}{(n_2 - x)!(n_3 - n_1 - 1)!} / \frac{n_2!(n_3 - 1)!}{(n_2 + n_3 - 1)!}. \tag{8.38}$$

This is easily seen to be

$$f(x; n_1, n_2, n_3) = \binom{n_1 + x - 1}{x} \binom{n_3 + n_2 - n_1 - x - 1}{n_2 - x} / \binom{n_3 + n_2 - 1}{n_2}, \tag{8.39}$$

for $x = 0, 1, \ldots, n_2$, which is the PMF of NHGD. By using $\binom{-n}{x} = (-1)^x \binom{n+x-1}{x}$, the PMF can be written alternatively as

$$f(x; a, b, n) = \binom{-a}{x} \binom{-b}{n-x} / \binom{-(a+b)}{n}, \tag{8.40}$$

[2]There are two popular beta distributions called type-I and type-II. Type-I beta has range $[0, 1]$, whereas type-II beta has range $[0, \infty)$.

for $x = 0, 1, 2, \ldots, n$ where a, b, n are integers. It can also be expressed in terms of generalized hypergeometric functions. When $a = b = 1$, this reduces to the rectangular distribution. An approximation to BBD in terms of negative binomial distribution appears in [103].

Analogous to the relationship between the CDF and SF of binomial and negative binomial distributions, a similar relationship exists among the CDF and SF of NHGD and HGD. Let $F(x; k, r, N)$ be the CDF of NHGD, $FD(x; k, r, N)$ be the CDF of DNHGD, and $G(x; k, r, N)$ be the CDF of HGD. Then $F(x; k, r, N) = 1 - G(r - 1; k, x, N)$, which is the SF of HGD. Similarly, $FD(x; k, r, N) = 1 - G(r - 1; N - k, x, N)$.

8.3.5   MEAN AND VARIANCE OF BBD

As $E(X) = E[E(X|p)]$, we get the mean as $\mu = nE(p) = na/(a + b) = nP$ where $P = a/(a + b)$. The second raw moment $\mu'_2 = \mu * [n(1 + a) + b]/(a + b + 1)$, from which the variance follows as $\sigma^2 = abn(a + b + n)/[(a + b)^2(a + b + 1)]$. As $Q = 1 - P = b/(a + b), \sigma^2 = nPQ(a + b + n)/(a + b + 1)$, which is obviously greater than the mean. Hence this distribution is overdispersed. In Bayesian analysis, this is written as $n\pi(1 - \pi)[1 + (n - 1)\rho]$ where $\pi = P$ and $\rho = 1/(a + b + 1)$ is the pairwise correlation between the trials called overdispersion parameter. This form is obtained from the previous one by writing a+b+n as $a + b + 1 + (n - 1)$ and dividing by the denominator $a + b + 1$.

In comparison with a binomial distribution with the same mean as BINO(n, P) where $P = a/(a + b)$ and variance $\sigma_b^2 = nP(1 - P) = nab/(a + b)^2$, we see that $\sigma^2 = \sigma_b^2(a + b + n)/(a + b + 1)$. As $(a + b + n)/(a + b + 1)$ is always $> 1, \sigma_b^2$ is less than the variance of BBD (equality holds for $n = 1$). This shows that the BBD is a suitable choice when looking for a distribution with the same mean as binomial distribution, but with larger variance.

See Table 8.2 for other properties. The distribution of waiting time until the kth success-run of fixed length in hypergeometric trials is also negative hypergeometric (Godbole (1990) [27], [69]).

8.3.6   MEAN DEVIATION

The mean deviation is most easily found using the MDGF (Chattamvelli and Shanmugam (2019) [18], pp. 43–45) as $D_x(t) =$

$$2E\left(t^x/(1 - t)^2\right) = 2t^r k_{(r)}/\left[N_{(r)}(1 - t)^2\right] {}_2F_1(r, -(N - k); -(N - r); t), \qquad (8.41)$$

as the coefficient of $t^{\lfloor \mu \rfloor - 1}$ where μ is the mean with a correction term added when the mean is not an integer. Alternately, it can be expressed in terms of $m = \lfloor \mu + 1 \rfloor$ as (Kamat (1965) [41])

$$\text{MD} = 2m(Nq - n + m)/N \binom{Np}{m}\binom{Nq}{n - m}\bigg/\binom{N}{n}. \qquad (8.42)$$

Table 8.2: Properties of beta-binomial distribution

Property	Expression	Comments
Range of X	$x = 0, 1,\ldots, n$	Discrete, finite
Mean	$\mu = na/(a + b) = n*P$	
Variance (σ^2)	$nPQ(a + b + n)/(a + b + 1)$	$\mu < \sigma^2$
Skewness γ_1	$(C + n)(b - a)/(D + 1)\sqrt{D}/[nabC]$	$C = a + b + n, D = a + b + 1$
CV	$\sqrt{b(a + b + n)/[na(a + b + 1)]}$	$\sqrt{bC/[naD]}$
MD	$2\mu_2 * f_m[1 + 1/(n + 1)]$	
$E[X\ldots(X - k + 1)]$	$k!/(1 - q)^{k+1} = k!/p^{k+1}$	
CDF	$[1 - B(n - k + b - 1, k + a + 1)_3F_2(a,b,k)]/K$	$K = [B(a,b)B(n - k, k + 2)(n + 1)$
PGFs	$\binom{r}{k} B(k + a, n - k + b)/B(a,b)$	
FMGF	$_2F_1[a, -n, a + b, -t]$	

8.4   TRUNCATED NHGD

The zero-truncated negative hypergeometric (ZTNHGD) distribution is used in many fields including CMR models. It is also used in validation studies that involve two independent processes.

The PMF is $f(x; p, N, n) =$

$$\binom{x + r - 1}{x}\binom{N - r - x}{k - r} / \left[[1 - \binom{N - r}{k - r} / \binom{N}{k}]\binom{N}{k} \right]$$
$$= \binom{x + r - 1}{x}\binom{N - r - x}{k - r} / \left[\binom{N}{k} - \binom{N - r}{k - r} \right],$$

(8.43)

where $x = 1, 2, \ldots, \min(n, k)$.

Left-truncated hypergeometric distribution is obtained by truncating at a positive integer K. A right-truncated hypergeometric distribution is also possible, but of limited utility.

Problem 8.2   Suppose missiles are fired from a ship thousands of miles away to hit a target. Each missile has probability p of hitting its designated target. What is the number of missile shots needed to have k hits? What is the distribution of $X =$ missed shots?

8.5   APPLICATIONS

This distribution is more popular in some fields than the positive hypergeometric distribution. Eliminating high-risk objects or entities from a group is often practiced in several fields like actuarial sciences, medical sciences, and engineering. As an example from actuarial sciences, suppose there are N insured persons or items (like automobiles or buildings). If k of them are considered to be high-risk, we may wish to eliminate r of them as inviable. Then X, the number of items selected to get r of high-risk category, is distributed as $NHGD(k, r, N)$. This model can be easily adapted to other fields. Suppose there are k out of N persons in a community who are of high-risk category to an epidemic like COVID-19. Assume that they are identifiable only through a screening test. If an isolation facility has a maximum capacity to accommodate r persons, we wish to conduct screening test to individuals one-by-one until r of them have been found. Then the number of persons selected (X) is distributed as $NHGD(k, r, N)$. Quantum cryptography uses it to model phase error rate in Quantum Key Distribution (QKD) protocols with a finite number of qubits. It is also used in contingency table analysis and queuing models.

8.5.1   COMPUTER GAMES

Consider a slot game in which a player is presented with N number of covered slots (usually generated randomly and arranged as a matrix with each cell representing a slot, or as some figure like balloons) which may contain awards (monetary reward) or losses (losses are called devils). A player is allowed to randomly select the slots one-by-one (by touching on the slot, or clicking on it using mouse, joystick, etc.) until r of the award slots are revealed (or its dual in which r devils or jokers are revealed). As the player does not know which are the covered slots with awards, they may select X other slots that do not carry awards (x can also be taken as the total number of slots selected as in the shifted NHGD). Assume that there are $k < N$ actual award slots. If selecting an award winning slot is considered as a success, and selecting devil slot as a failure, the problem can be formulated as $NHGD(k, r, N)$ because it is like sampling without replacement. The expected amount can then be calculated using PMF for x number of correctly revealed slots.

8.5.2   CMR ESTIMATION

The last chapter discussed the CMR model in detail. Suppose a population size is unknown, but an upper-bound N is available. A small fraction of the population (called marking sample of size k) is selected. Each item is marked and released as mentioned earlier. Then in the recapture phase, items are selected at random without replacement from the population until r ($< k$) of the marked items have been selected (captured). Let X denote the number of items selected. Then X has an $NHGD(k, r, N)$.

8.5.3   INDUSTRIAL QUALITY CONTROL

It is used in a product manufacturing environment, when either the testing is expensive or destructive. Destructive testing involves products or parts that cannot be shipped because the testing process itself makes it unusable. In such a scenario, it is the usual practice to select items one by one from a lot of size N until a certain number r of defectives (that do not meet quality specifications) have been found. The proportion of defectives items are usually known from prior data. If X denotes the number of such selected items, p denotes the proportion of defectives, the distribution of X can be expressed as IHD. The lot is rejected if the number of defectives exceed a cutoff c.

8.6 SUMMARY

The negative hypergeometric and beta-binomial distributions are introduced in this chapter. Various representations of the PMF are mentioned, and important properties are discussed. Inter-relationships among various distributions are discussed.

CHAPTER 9

Logarithmic Series Distribution

After finishing the chapter, readers will be able to ...

- Understand logarithmic series distribution.

- Describe alternate forms of logarithmic series distribution and their properties.

- Explore its relation with negative binomial distribution.

- Apply logarithmic series distribution in ecology, finance, etc.

9.1 INTRODUCTION

Consider the expansion $(1-q)^{-1} = 1/(1-q) = 1 + q + q^2 + q^3 + \cdots$. Integrate both sides w.r.t. q to get $-\log(1-q) = q + q^2/2 + q^3/3 + \cdots$, where $\log()$ denotes logarithm to the base e, which is sometimes denoted as $\mathtt{ln}$ (we will stick with $\mathtt{log}$ notation because logarithm to other bases are not discussed). Put $1 - q = p$, and divide both sides by $-\log(p)$ to get the logarithmic series distribution, denoted by $\mathrm{LSD}(p)$. Ecology and biological sciences use θ, and some engineering fields use π instead of p. Although $\log(x)$ is a continuous function, this is a discrete distribution with infinite support. It has PMF

$$f(x, p) = \begin{cases} q^x/[-x \log p] = (1-p)^x/[-x \log p] & \text{for} \quad 0 < p < 1, \quad x = 1, 2, \ldots \\ 0 & \text{otherwise.} \end{cases}$$

This is also known as log-series or logarithmic distribution. It belongs to the power series distribution as it can be obtained using McLaurin series expansion of $-\log(1-q)$.

9.1.1 ALTERNATE FORMS

Swapping the roles of p and q results in the PMF $f(x, p) = p^x/[-x \log(q)]$, $x = 1, 2, \ldots \infty$. This is called the dual logarithmic-series distribution. All properties are obtained by swapping p and q. An alternate parametrization is obtained by putting $q = 1 - p = \theta$ as

$$f(x, \theta) = \begin{cases} \alpha \, \theta^x/x & \text{for} \quad \alpha = -[\log(1-\theta)]^{-1}, \; 0 < \theta < 1, \quad x = 1, 2, \ldots \\ 0 & \text{elsewhere.} \end{cases}$$

Put $1/p = \lambda$ (so that $0 < \lambda < \infty$) to get another form as

$$f(x, \lambda) = \begin{cases} (1 - 1/\lambda)^x/[x \log(\lambda)] & \text{for} \quad \lambda = 1/p, \quad x = 1, 2, \ldots \\ 0 & \text{elsewhere.} \end{cases}$$

Then $p \to 0$ is equivalent to $\lambda \to \infty$.

Put $-\log(q) = \lambda$ to get

$$f(x, \lambda) = \begin{cases} \exp(-\lambda x)/[x \log(1 - \exp(-\lambda))] & \text{for} \quad 0 < \lambda < \infty, \quad x = 1, 2, \ldots \\ 0 & \text{elsewhere.} \end{cases}$$

Next put $\lambda = q/p$ to get

$$f(x; \lambda) = (\lambda/(1 + \lambda))^x/[x \log(1 + \lambda)], \quad x = 1, 2, \ldots. \tag{9.1}$$

A shifted version has PMF

$$\begin{aligned} f(x; p) &= q^{x+1}/[-(x + 1) \log p] \\ &= (1 - p)^{x+1}/[-(x + 1) \log p] \quad \text{for} \quad 0 < p < 1, \quad x = 0, 1, 2, \ldots. \end{aligned} \tag{9.2}$$

9.1.2   RELATION TO OTHER DISTRIBUTIONS

This distribution approaches $GEO(p)$ when $p \to 0.50$, and the degenerate distribution at $x = 1$ when $p \to 1$ (say $p > 0.95$). This is a special case of the left-truncated negative binomial distribution where the zero class has been omitted, and the parameter k tends to one. Anscombe (1950) [2] represented the PMF of negative binomial distribution in terms of gamma function as $f(x; k, n, p) = \Gamma(k + n)/(n!\Gamma(k)) \, q^k p^n$, and obtained LSD as $n \to \infty$, and $k \to 0$. He applied it to multiple species sampling from an unknown number of species, as also Jain and Gupta (1973) [36]. The same technique is applied in medical sciences to study a variety of diseases (like influenza or COVID like diseases in which viruses undergo genetic mutations resulting in multiple sub-types). Ord (1967) [60] used graphical methods to prove that negative binomial distribution can be approximated by LSD when $q \to 0$ (meaning that the number of failures become negligibly small as n becomes large). The Ewens sampling formula encountered in genomics that provides a null-hypothesis distribution of allele frequencies of n genes into allelic types uses a multivariate extension of LSD. If $X_1, X_2, \ldots, X_N$ are IID logarithmic variates with parameter p, where $N \sim POIS(\lambda)$ then $Y = X_1 + X_2 + \cdots + X_N = \sum_{j=1}^{N} X_j$ has a negative binomial distribution.

The distribution of the sum of a fixed number of IID $LSD(p)$ has a Stirling distribution of first kind (SDFK). This is most easily proved by MGF or PGF. As they are IID, the PGF of $Y = X_1 + X_2 + \cdots + X_n$ is the product of the PGFs. Thus, $P_y(t) = [\log(1 - qt)/\log(p)]^n$,

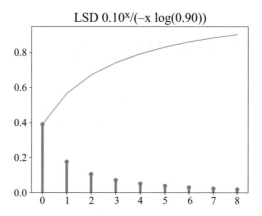

Figure 9.1: PMF and CDF of LSD (0.10).

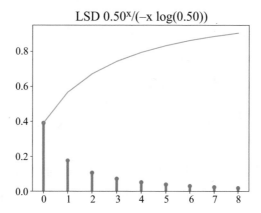

Figure 9.2: Logarithmic distribution ($p = 0.5$).

for $|t| < 1/q$. Hence,

$$f(y; n, p) = \Pr[Y = y] = \Gamma(n + 1)|s(x, n)|q^x/[x!(-\log(p))^n], \quad \text{for } x = n, n + 1, \ldots,$$
$$(9.3)$$

where $s(x, n)$ are Stirling number of first kind (SNFK). In particular, the statistic $(X_1 + X_2 + \cdots + X_n)/n$ tends to the Stirling distribution when n increases, and will eventually tend to normality as $n \to \infty$. As the nth derivative of $-\log(1 - ax)$ is $n!a^n/(1 - ax)^n$, the negative binomial and geometric distributions can be obtained by differentiating $-\log(1 - q)$ w.r.t. q (Sadinle M. (2008) [70]). It is also related to the Poisson distribution as $p \to 0$ (Quenouille (1949) [65]). Shanmugam and Singh (1995) [94, 95] obtained a weighted LSD using a convolution $Y = X + Z$ where $X \sim \text{LSD}(\theta)$, and $Y \sim \text{NBINO}(r, \theta)$ useful in purchasing behavior

Table 9.1: Properties of logarithmic distribution

Property	Expression	Comments
Range of X	$x = 1,\dots,\infty$	Discrete, infinite
Mean	$\mu = q/[-p\log(p)]$	$\alpha\theta/(1-\theta)$
Variance σ^2	$-q(q+\log(p))/(p\log(p))^2 = \mu(1/p - \mu)$	$= \mu((1-\theta)^{-1} - \mu, \Rightarrow \mu < \sigma^2$
Mode	$1/\log(\theta)$	
Skewness	$\gamma_1 = [(1+\theta)-3D+2D^2]/[\sqrt{D}(1-D)^{3/2}$	$D = \alpha\theta$
Kurtosis	$(1+4\theta+\theta^2-4D(1+\theta)+6D^2-3D^3/R$	$R = D(1-D)^2$
CV	$\sqrt{(1-\alpha\theta)}/\sqrt{\alpha\theta}\,(1+\theta)^2$	$(\alpha^{-1}\theta^{-1})^{1/2}$
Mean deviation	$2a\sum_{k=0}^{\lfloor\mu\rfloor}(\mu-a)/a\theta^a$	
Mom. recurrence	$\mu'_{r+1} = \theta[\partial/\partial\theta + \alpha/(1-\theta)\,\mu'_r$	
Mom. recurrence	$\mu_{r+1} = \theta\partial\,\mu_r/\partial\theta + r\mu_2\,\mu_{r-1}$	
Factorial moments	$\alpha\theta^r\,(r-1)!/(1-\theta)^r$	$(r-1)!(q/p)^r/-\log(p)$
PGF	$\log(1-\theta t)/\log(1-\theta)$	
MGF	$\log(1-\theta\exp(t))/\log(1-\theta)$	
Recurrence	$f(x+1;p)/f(x;p) = qx/(x+1)$	$(1-p)(1-\frac{1}{x+1})$
Variance is less than the mean for $(1/p-\mu)<1$ or equivalently $p>1/(1+\mu)$.		

modeling. See Khang and Ong (2007) [45] for a generalization of LSD using inverse trinomial distribution.

9.2   PROPERTIES OF LOGARITHMIC DISTRIBUTION

The logarithmic distribution has a single parameter p. Due to the presence of x in the denominator, a change of origin transformation $Y = X - c$ (for $c > 0$) is not meaningful, but $Y = X + c$ (for $c > 0$) that shifts the distribution to the right is permitted. A recurrence relation for the PMF follows easily as $f(x+1;p)/f(x;p) = xq/(x+1)$, which is a decreasing function in x. Hence the mode is $x = 1$.

9.2.1   MEAN AND VARIANCE

Moments can be found either directly or using the following theorem.

Theorem 9.1   *The kth moment of LSD is C times (k-1)th moment, and kth inverse-moment is C times (k+1)th inverse-moment of ZTG distribution where $C = -q/[p\log(p)]$.*

Proof. Consider the kth moment as

$$\mu_k = -1/\log(p) \sum_{x=1}^{\infty} x^k q^x / x = -q/[p \log(p)] \sum_{x=1}^{\infty} x^{k-1} q^{x-1} p. \qquad (9.4)$$

The summation term is the $(k-1)$th moment of zero-truncated geometric distribution for $k > 1$, and it evaluates to 1 (being the sum of all probabilities) for $k = 1$. Hence, the mean is $\mu = -q/[p \log(p)]$. As $0 < p < 1$, the $\log(p)$ term is negative, thereby the negative sign cancels out giving the mean as positive. A new representation follows from this as

$$f(x, p) = \begin{cases} \mu q^{x-1} p/x & \text{for } 0 < p < 1, \ x = 1, 2, \ldots \\ 0 & \text{elsewhere.} \end{cases}$$

This shows that the tailing-off is faster than for the geometric law. The above expression can be used to find the first inverse moment of truncated geometric distribution easily.

From (9.2), $E(X^2)$ is the same as $-q/[p \log(p)] \times$ mean of ZTGD GEO(p), which from Chapter 4 is $1/p$. This gives $E(X^2) = -q/[p \log(p)] \times 1/p = -q/[p^2 \log(p)]$. Thus, the variance is $\sigma^2 = -q(q + \log(p))/[(p \log(p))^2]$. In terms of μ this is $\sigma^2 = \mu(1/p - \mu)$ or equivalently $\sigma^2 + \mu^2 = \mu/p$. Cross-multiply to get $p = \mu/(\sigma^2 + \mu^2)$. This shows that the variance is less than the mean for $(1/p - \mu) < 1$ or equivalently $p > 1/(1 + \mu)$.

For the alternate form, the mean is $\theta/[(\theta - 1)\log(1 - \theta)]$ and variance is $\mu(1/(1 - \theta) - \mu)$ (Table 9.1). The CV is $\sigma/\mu = -q/[p \log(p)]$.

The IJR is undefined as x does not take the value 0, but $\Pr[X = 2]/\Pr[X = 1]$ is $q/2$. For the alternate representation above, the mean is $\mu = a\theta/(1 - \theta)$ and variance is $\mu(1 - a\theta)/(1 - \theta)$. To fit the model, compute $\bar{x}$ and s^2, and find $\hat{p} = \bar{x}/(\bar{x}^2 + s^2)$. As the variance is $+ve$, $q < -\log(p)$. $\qquad \square$

9.2.2 FACTORIAL MOMENTS

The factorial moments are easier to find than central moments. The kth factorial moment is given by $\mu_{(k)} = E[x(x - 1) \ldots (x - k + 1)] =$

$$-\log(p) q^k \partial^{k-1}/\partial q^{k-1} \left[\sum_{j=1}^{\infty} q^{j-1} \right] = -\frac{(k-1)!}{\log(p)} (q/p)^k. \qquad (9.5)$$

The factorial moments can also be found from that of geometric distribution

$$\mu_{(k)} = E[x(x - 1) \ldots (x - k + 1)] = cq^k \sum_{x=1}^{\infty} x(x - 1) \ldots (x - k + 1) q^x / x \qquad (9.6)$$

$$= cq/p \sum_{x=2}^{\infty} (x - 1) \ldots (x - k + 1) q^{x-1} p, \qquad (9.7)$$

where $c = -\log(p)$. It is easy to note that the sum is the factorial moment of $GEO(p)$.

9.2.3   CDF AND SF

This has CDF $c \sum_{j=1}^{\lceil x \rceil} q^j/j$ for the first version, and $\alpha \sum_{j=1}^{\lceil x \rceil} \theta^x/x$ for alternate version. The CDF in terms of incomplete beta function is $1 + I_q(x + 1; 0)/\log(p)$.

Example 9.2   SF of logarithmic distribution   Prove that the survival function is bounded by $[(1 - q/(x + 1))^{-1} f(x; p) < SF < (1 - q/x)^{-1} f(x; p)$.

Solution: By definition SF is $S(x) = c \sum_{x=k}^{\infty} q^x/x$ (we sum from k onwards instead of $k + 1$). Write the sum in expanded form

$$q^x/x + q^{x+1}/(x + 1) + q^{x+2}/(x + 2) + \cdots < q^x/x + q^{x+1}/x^2 + q^{x+2}/x^3 + \cdots . \quad (9.8)$$

Take q^x/x as common factor to get $S(x) < q^x/x[1 + q/x + q^2/x^2 + \cdots] = q^x/x[1 - q/x]^{-1}$. Now multiply by c and identify $f(x; p) = cq^x/x$. Next consider $S(x) = c \sum_{x=k}^{\infty} q^x/x > q^x/(x + 1) + q^{x+1}/(x + 1)^2 + q^{x+2}/(x + 1)^3 + \cdots$. Take $q^x/(x + 1)$ as common factor to get $S(x) > cq^x/(x + 1)[1 + q/(x + 1) + q^2/(x + 1)^2 + \cdots] = cq^x/(x + 1)[1 - q/(x + 1)]^{-1}$. Now identify $f(x; p) = cq^x/x$ to get the result.

9.2.4   GENERATING FUNCTIONS

The PGF is $P_x(t; p) =$

$$-(1/\log(p)) \sum_{x=1}^{\infty} t^x q^x/x = (1/\log(p)) \log(1 - qt) = \log(1 - qt)/\log(1 - q), \quad (9.9)$$

for $|t| < 1/q$. The MGF follows as $M_x(t; p) =$

$$-(1/\log(p)) \sum_{x=1}^{\infty} \exp(tx)q^x/x = \log(1 - q\exp(t))/\log(p) = \log(1 - q\exp(t))/\log(1 - q),$$

$$(9.10)$$

Take log of both sides to get the cumulant CGF as

$$K_x(t; p) = \log(M_x(t; p)) = \log(\{\log(1 - q\exp(t))/\log(1 - q)\}). \quad (9.11)$$

From the PGF, the CDFGF follows easily as $\log(1 - qt)/[\log(p)(1 - t)]$.
The factorial MGF is

$$E[(1 + t)^x] = c \sum_{x=1}^{\infty} (1 + t)^x q^x/x = c \sum_{x=1}^{\infty} [q(1 + t)]^x/x$$

$$(9.12)$$

$$= -\log(1 - q(1 + t))/\log(p).$$

This could also be written as

$$E[(1 + t)^x] = -\log(p - qt))/\log(p).\qquad(9.13)$$

The ChF is given by $\phi(t) = \ln(1 - qe^{it})/\ln(1 - q)$. As the values assumed by X are integers, it is used in those modeling situations involving counts. For instance, the number of items of a product purchased by a customer in a given period of time can be modeled by this distribution. See Table 9.1 for more properties.

The MDGF is easily obtained as $2\log(1 - qt)/[\log(p)(1 - t)^2]$. When the mean is an integer, the MD is the coefficient of $t^{\mu-1}$ in the power series expansion of MDGF. A correction term mentioned in Chapter 1 must be added when μ is non-integer.

Example 9.3   Logarithmic probabilities   If $X \sim \text{LSD}(p)$, find the following probabilities: (i) X takes even values, and (ii) X takes odd values.

Solution: As the logarithmic distribution takes $x = 1, 2, \ldots, \infty$ values, both the above probabilities are evaluated as infinite sums. (i) $P[X \text{ is even}] = c[\log(1 - qt) + \log(1 + qt)]/2 = (c/2)(\log[(1 - qt)(1 + qt)]) = (c/2)\log[(1 - q^2t^2)]$. (ii) $P[X \text{ is odd}] = c[\log(1 - qt) - \log(1 + qt)]/2 = (c/2)(\log[(1 - qt)/(1 + qt)])$, where $c = -1/\log(p)$.

Example 9.4   Expected value of $(-1)^x$ If $X \sim \text{LSD}(p)$, find expected value of $(-1)^x$.

Solution: By   definition, $E[(-1)^x] = \sum_{x=1}^{\infty}(-1)^x q^x/[-x\log(p)] = -1/\log(p)$ $\sum_{x=1}^{\infty}(-q)^x/x = -\log(1 + q)/\log(p)$.

Example 9.5   Expected value of e^{-x}   If $X \sim \text{LSD}(p)$, find expected value of e^{-x}.

Solution: By definition $E[e^{-x}] = \sum_{x=1}^{\infty} e^{-x}q^x/[-x\log(p)]$. Take constants outside the summation to get $-(1/\log(p))\sum_{x=1}^{\infty}(q/e)^x/x = -\log(1 - q/e)/\log(p)$.

Problem 9.6   Prove that the sum of N logarithmic distributions where N itself has a Poisson distribution results in a negative binomial distribution.

A characterization of LSD appears in Shanmugam and Singh (1984).[1] If $X_1, X_2, \ldots, X_n$ are IID $\text{LSD}(p)$ variates, then for any positive integer $k(\geq n)$, and $Y = X_1 + X_2 + \cdots + X_n = \sum_{j=1}^{n} X_j$, $\Pr[Y = k] = c^n q^k \sum_{Y=k} \prod_{j=1}^{n}(1/x_j) = c^n q^k n!|s(k, n)|/k!$, for $k \geq n$, where $|s(k, n)|$ are unsigned SNFK. As the conditional distribution of $X_i = x|\sum_{j=1}^{n} X_j = k$ is

$$\Pr\left[X_i = x \Big| \sum_{j=1}^{n} X_j = k\right] = \binom{k}{x} F(x, 1)F(k - x, n - 1)/(nF(k, n)),\qquad(9.14)$$

---

[1]These authors used the index variable i twice in their expression given above.

which is independent of q, the statistic $t = \sum_{j=1}^{n} X_j$ is unbiased, complete, and sufficient for q. See Shanmugam and Singh (1984) [94] for a test statistic to check whether a random sample has been drawn from an LSD(p) distribution.

9.3   ZERO-INFLATED LOGARITHMIC DISTRIBUTION

As the logarithmic distribution has domain $1, 2, \ldots, \infty$, it cannot be used to model count data containing zeros. A solution is zero-inflated logarithmic series distribution (ZILSD) with PMF

$$g(x, p) = \begin{cases} 1 - \pi & \text{for } x = 0 \\ \pi q^x / [-x \log p] & \text{for } 0 < p < 1, \ x = 1, 2, \ldots \end{cases}$$

which increases the probability of zero count. This model is used in many fields like ecology and finance (discussed below).

9.4   LAGRANGIAN LOGARITHMIC DISTRIBUTION

Several generalized forms of LSD are available. Consider the PMF

$$f(x; k, p) = C \frac{1}{kx} \binom{kx}{x} p^x (1 - p)^{kx-x}, \ x = 0, 1, \ldots \tag{9.15}$$

for $10 \le k \le 1/p$, and $C = -1/\log(1 - p)$. This is called Lagrangian LSD.

A three parameter extension of the logarithmic distribution was given by Shanmugam (1988) [76] with PMF

$$f(x; c, b, p) = c[\Gamma(bx)/\Gamma(bx - x + 1)](g(q))^x / x!, \quad \text{for } \ x = 1, 2, 3, \ldots, \infty, \tag{9.16}$$

where $c = -1/\log(p)$ and $g(q) = q(1 - q)^{b-1}$. This reduces to the LSD(p) when $b = 1$.

9.5   APPLICATIONS

The LSD has gained popularity among actuarial and related sciences. For example, insurance claims frequency are modeled using LSD (or its zero-inflated version when zero counts are to be included). In medical sciences it is used to model the number of parasites per host, dental caries filled, number of medical procedures (like x-rays, CAT scans) undergone in a fixed time period (say one decade), etc. It is also used to model patient specific medical conditions (usually in geriatrics) like number of seizures, premature ventricular contractions in arrhythmic patients, etc. The zero-truncated version is used to model number of patient visits to specialist doctors, clinics, hospitals, etc. Credit card companies use it to model payment defaulters, overdrafts, etc. Retail industry uses it to model number of items purchased by customers (like fans, memory cards, clothes, soxes, etc.) or student purchases (notebooks, pen, shirts) in a particular time period. It is also used by demographers to model count of households or counties as per the number of migrant workers. Government agencies use it to model unemployment data, social security benefits and insurance data. Meteorologists use it to model dry and wet runs of days, say, in rainy season. If k number of sunny days occur together, we have a dry run of length k. Frequency distribution of runs of length k need not form an LSD for all regions, but provides an approximate fit for some regions (especially tropical land) during monsoon season. Similarly criminologists use it to model runs of days (say 24-hour durations) with k major crimes in large cities.

It is also used in several engineering fields. Consider repairable devices and gadgets like induction cookers, smart devices, etc. We could build models for individual gadgets or brands using available repair data with LSD (or ZTLSD). In reliability engineering the standard model of a systems consists of n components, each working independently of the others. Then the failures in parallel circuits over a period of time can be modeled using LSD. Some manufacturing industries use it to model the total number of product defects. Highway engineers use it to model the number of days n with x accidents where $x = 0, 1, 2, \ldots$ for the zero-inflated model, and $x = 1, 2, \ldots$ (no accidents event is dropped) for the standard model.

9.5.1   SCIENTIFIC RESEARCH

Consider the problem of modeling the number of papers published by researchers in a field during a fixed time interval (say one year). Assume that those who didn't publish during this time are excluded from the study. Then let f1 denote the number of persons (count) who have just one publication, f2 denote the number of persons who have two publications, and so on. Then we could fit an LSD for large samples. A discontinuity problem may arise for small samples. For example if among 30 faculty members in a department, several have 2 and 4 publications, but none have 3 publications, we have a discontinuity at $x = 3$. This kind of problem is rare for large samples although it could still occur toward extreme right tail (say none published 13 papers, but few have 14 or more publications). A right-truncated LSD may be appropriate in such cases if this type of discontinuity occurs in extreme right tail.

9.5.2   ECOLOGY

Several applications of LSD exist in ecology [66]. Fisher et al. (1943) [25] used it to model relative species abundance (of butterflies) in the Malayan peninsula, and of moth species (in UK). Relative species abundance is a biodiversity measure of how rare or common a species in an enclosed region or community is relative to other species in the region (usually at single trophic level or single phylum like birds, crabs, butterflies, moth, etc.) [105]. Probability of a population of size n is typically proportional to $1/n$, which suggests an LSD to model it. Species richness denotes the number of genetically different species in a community. Marine biologists use it to model moving creatures at specific locations. It is also used to model parasitic infestations in marine creatures like crabs, fishes, etc. These conform to specific patterns that are of interest in macroecology. They are usually plotted on logarithmic scale (log of abundance bins on X-axis, number of species on Y-axis).

It can be used to model many other count data like number of species in enclosed area, birds sighted in a short time period, plant species in forest areas, aquatic life in reservoirs, parasites on the body surface of creatures, etc.

9.5.3   FINANCE

The rate of returns in financial market varies on each trading day on which the stocks are traded. Without loss of generality, we count the loss or gain at the end of a trading day. It may be convenient to reckon this in local currency in each country where stocks are held. Unit intervals are then defined using the currency (say in increments of \$5 as \$1-5, \$5-10, etc.) Separate loss model and gain model can now be developed depending on the frequency distribution of trading days matched with above defined intervals. A ZILSD (described below) may be used when no loss or gain are observed on some trading days. Appropriate selection of unit interval is crucial in getting a good fit of LSD.

Stock assets are usually combined into portfolios. A portfolio comprises of multiple well-trading stocks. An LSD model can be developed for each such portfolio or all portfolios held by a customer.

9.6   SUMMARY

The classical logarithmic series distribution is introduced in this chapter, and its basic properties are derived. Its connection to other distributions are briefly discussed, including zero-truncated and zero-inflated logarithmic series distributions. Some applications in ecology and scientific research are also mentioned.

CHAPTER 10

Multinomial Distribution

After finishing the chapter, readers will be able to . . .

- Understand multinomial theorem and its forms.

- Explain multinomial trials and multinomial coefficients.

- Describe multinomial distribution and its properties.

- Apply multinomial distribution in practical situations.

10.1 INTRODUCTION

This distribution can be considered as a generalization of the binomial distribution with $k(>2)$ categories. It gives the joint probability of occurrence of multiple events in n independent trials. Practical experiments that result in a multinomial distribution can be characterized by the following properties.

1. The experiment consists of a series of IID trials with two or more (say k) mutually exclusive categorical outcomes. Each categorical event is assumed to be independent.

2. The trials can be repeated independently n times under identical conditions. The outcome of one trial has no effect on the outcome of any other, including next trial.

3. The probability of occurrence of one of the outcomes p_k remains the same from trial to trial until the experiment is over.

4. The random variable X denotes the number of times outcome X_k takes the value x_k.

Thus, it describes the theoretical distribution of n objects selected at random from a categorical population with more than two categories. Examples are election results for multiple political parties, spread of epidemics (or death due to them) among different ethnic groups, inclusion of stocks in portfolios, accidents due to different vehicle types (cars, buses, trucks, etc.), claims received under various types in actuarial sciences, books of various types (like fiction, novels, stories, poems, science) checked-out by patrons, etc. Geologists use it for soil analysis, and bacteriologists use it to model microbe counts by type in randomly distributed colonies. Likewise, civil engineers design structures in such a way as to withstand rare events like earthquakes, floods, high winds, tornadoes, fires, etc. These of course depend on the location of the structure. If the respective probabilities are known from prior data, a multinomial distribution can be used to estimate the probability of structural damage due to multiple causes (e.g., earthquake and fire) over a time period (say one week). This distribution arises in practice while sampling from a finite population with replacement, or an infinite population. It is assumed in both cases that there are k distinct categories in the population, the probabilities of which are known. In engineering and applied sciences this means that each experiment can result in any one of the k possible outcomes.

Let the probabilities be denoted by p_i for the ith class such that $\sum_{i=1}^{k} p_i = 1$. We denote the distribution by $MN(n, p_1, p_2, \ldots, p_k)$ or $MN(n, k, p_1, p_2, \ldots, p_k)$ (where k is the number of categories, which is implicit in the first notation as the number of p's). As MN could be confused with "multivariate normal," an alternate notation is $MULT(n, k, p_1, p_2, \ldots, p_k)$. The entire vector of p's (of size k) can be represented as $\mathbf{P}_k$, and the above represented as $MN(n, \mathbf{P}_k)$ or $MULT(n, \Pi)$.

The PMF of a general multinomial distribution with k classes is

$$f(x; n, p_1, p_2, \ldots, p_k) = \begin{cases} \frac{n!}{x_1!x_2!\ldots x_k!} p_1^{x_1} p_2^{x_2} \cdots p_k^{x_k} & \text{for } x_i = 0, 1, \ldots, n \\ 0 & \text{elsewhere,} \end{cases}$$

where $x_1 + x_2 + \cdots + x_k = n$ and $p_1 + p_2 + \cdots + p_k = 1$. It is so-called because it can be obtained from the expansion of $(p_1 + p_2 + \cdots + p_k)^n = \sum (n! / \prod_{j=1}^{k} n_j!) \prod_{j=1}^{k} p_j^{n_j}$. Each of the x_k could range from 0 to n inclusive. Due to the linear constraint, only $(k - 1)$ variables are involved.

Problem 10.1   Pizza hut offers pizzas in four sizes, namely, small (S), medium (M), large (L), and extra large (X). It is known from past data that the respective order probabilities are 45%, 35%, 15%, and 5%. Find the probability of getting 110 S, 100 M, 30 L, and 10 X orders on a weekend.

10.1.1   ALTERNATE FORMS

The joint distribution of $X_1, X_2, \ldots, X_k$ is given by the multinomial coefficient $(p_1 + p_2 + \cdots + p_k)^n$. Using the product notation, the PMF can be written as $(n! / \prod_{i=1}^{k} x_i!) * \prod_{i=1}^{k} p_i^{x_i}$. This can also be written as $\binom{n}{x_1, x_2, \ldots, x_k}$, which is called the multinomial coefficient, and the variables $X_1, X_2, \ldots, X_k$ are multinomial variables. Some fields use π_i or θ_i to denote the probabilities, which are called cell probabilities. The parameter n (called size or index) is quite often an integer, but an extension to non-integers is possible because the multinomial coefficient is defined even for fractions. As $x_1 + x_2 + \cdots + x_k = n$, the PMF can alternately be written as

$$\frac{n!}{x_1!x_2!\ldots x_{k-1}!(n - \sum_{j=1}^{k-1} x_j)!} p_1^{x_1} p_2^{x_2} \cdots p_{k-1}^{x_{k-1}} (1 - p_1 - p_2 - \cdots p_{k-1})^{(n-\sum_{j=1}^{k-1} x_j)}, \quad (10.1)$$

where $\sum_{j=1}^{k-1} p_j < 1$ (here $(1 - p_1 - p_2 - \cdots, p_{k-1})$ is like the "q" of Binomial distribution). This can also be written as

$$\binom{n}{x_1, x_2, \ldots, x_{k-1}, n - \sum_{j=1}^{k-1} x_j} \prod_{i=1}^{k-1} p_i^{x_i} (1 - p_1 - p_2 - \cdots p_{k-1})^{(n - \sum_{j=1}^{k-1} x_j)}. \qquad (10.2)$$

This distribution can also be extended to non-integer values of n by replacing $n!$ by $\Gamma(n + 1)$. Similarly, it can be cast in the urn-model by throwing n independent balls into k independent urns, and counting the number of balls in the urns [38]. A special case occurs when all the probabilities are equal (say $1/k$; called equiprobability) for which simplified expressions ensue for most of the properties. For instance, the PGF becomes $(t_1 + t_2 + \cdots + t_k)^n / k^n$. A null statistical hypothesis of equiprobability ($H_0 : p_1 = p_2 = \cdots = p_k$) against the alternative that at least one of them is different results in the reduced model $\binom{n}{x_1, x_2, \ldots, x_k} p^n$ where $p = 1/k$, for which χ^2 tests are applicable when $n > 5k$. As shown below, all marginal distributions are BINO(n, p) in this case. The distribution of extremes and range of order statistics are easy to derive in the equiprobability case [9, 67].

10.2   RELATION TO OTHER DISTRIBUTIONS

If the number of items selected (n) is one, and k is two, we have a Bernoulli distribution. For $n > 1$ and $k = 2$, this reduces to BINO(n, p). The marginal distributions are also BINO(n, p_j) as shown below. It is called the trinomial distribution for $k = 3$. As the binomial distribution tends to the normal law as $n \to \infty$, the multinomial distribution tends to multivariate normal distribution asymptotically. As in the case of binomial distribution, we could show that this distribution tends to the multivariate Poisson distribution (see below).

It was mentioned in Chapter 7 that swapping the roles of probabilities of success and failures (p and q) in the hypergeometric distribution results in dual hypergeometric distribution. Suppose there are k items of first kind (success type) and $N - k$ items of second kind (failure type). If x items of the first type and $n - x$ items of the second type are selected from the population, there remains $k - x$ unselected items of the first type, and $N - k - (n - x) = (N - n - k + x)$ items of the second type in the population. Thus, the selection induces a partition of the population into four categories (selected and unselected of each type). This can be represented using the multinomial distribution as

$$f(x; k, N, n) = \binom{N}{x . n - x, k - x, N - n - k + x} / \left[\binom{N}{n} \binom{N}{k} \right]. \qquad (10.3)$$

Expand the multinomial coefficient and cancel out one $N!$ to get $f(x; k, N, n) =$

$$k!/[x!(k - x)!] \ (N - k)!/[(n - x)!(N - n - k + x)!] \ /[N!/(n!(N - n)!)]. \qquad (10.4)$$

This is the hypergeometric distribution with PMF $\binom{k}{x}\binom{N-k}{n-x}/\binom{N}{n}$ (Chapter 7, pp.136).

Table 10.1: Properties of multinomial distribution

Property	Expression	Comments
Range of X	$x_i = 0, 1, \ldots, n$	Discrete, $\sum_{i=1}^{k} x_i = n$
Mean	$\mu = np_i$	Need not be integer
Variance	$\sigma^2 = np_i q_i$	$\mu > \sigma^2$
Covariance	$-np_i p_j$	$i \neq j$
Mode	$(x-1), x[(n+1)*p_i$ not int]	$x = (n+1)*p_i$ else
Skewness	$\gamma_1 = [(1 - 2p_i)/\sqrt{np_i q_i}$	$= (q_i - p_i)/\sqrt{np_i q_i}$
Kurtosis	$\beta_2 = 3 + (1 - 6pq)/npq$	
$E[X(X-1)\ldots(X-r+1)]$	$n^{(\sum_{j=1}^{k} r_j)} \prod_{j=1}^{k} p_j^{r_j}$	
PGF	$[\sum_{j=1}^{k} p_j t_j]^n$	$p^n (1+t)^n$ if each $p_i = p$
MGF	$[\sum_{j=1}^{k} p_j e^{t_j}]^n$	
ChF	$\phi(t) = [\sum_{j=1}^{k} p_j e^{it_j}]^n$	

Never symmetric, always leptokurtic. Satisfies the additivity property $\sum_{i=1}^{m}$ MN$(n_i, p_1, p_2, \ldots, p_k)$ = MN$(\sum_{i=1}^{m} n_i, p_1, p_2, \ldots, p_k)$, if they are independent.

10.3 PROPERTIES OF MULTINOMIAL DISTRIBUTION

As the p_i's are constrained as $\sum_{i=1}^{k} p_i = 1$, there are k parameters (including n). As in the case of binomial distribution, the multinomial distribution is additive when all probabilities are equal. Symbolically, if $X \sim$ MN$(n_1, p_1, p_2, \ldots, p_k)$ and $Y \sim$ MN$(n_2, p_1, p_2, \ldots, p_k)$ then $X + Y \sim$ MN$(n_1 + n_2, p_1, p_2, \ldots, p_k)$. This result can be extended to any number of IID random variables. Analogous to the relationship between tail probabilities of binomial distribution and the incomplete beta function, there exist a similar relation between tail probabilities of multinomial distribution and the incomplete Dirichlet integrals. See Table 10.1 for more properties.

10.3.1 MARGINAL DISTRIBUTIONS

The marginal distributions are binomial, which follows easily from the observation that the probabilities are obtained as terms in the expansion of $(p_1 + p_2 + \cdots + p_k)^n$. If marginal distribution of x_j is needed, put $p_j = p$, and the rest of the sum as $1 - p$ (as $\sum_{i=1}^{k} p_i = 1, 1 - p = \sum_{i \neq j=1}^{k} p_i$). This results in the PGF of a binomial distribution. The marginal distributions can also be obtained for a fixed category (say x_j) directly by summing over other variables using

$$f(x_1) = \sum_{x_2, x_3, \ldots \in \Omega} f(x_1, x_2, \ldots, x_k), \tag{10.5}$$

$$f(x_1, x_2) = \sum_{x_3, x_4, \ldots \in \Omega} f(x_1, x_2, \ldots, x_k), \tag{10.6}$$

and so on. Without loss of generality, we will derive the marginal distribution of x_1. This is obtained by summing over other variables as $f(x_1) = \Pr[X_1 = x_1] =$

$$p_1^{x_1} \sum_{x_2 + x_3 + \cdots + x_k = n - x_1} \binom{n}{x_1, x_2, \ldots, x_k} p_2^{x_2} p_3^{x_3} \cdots p_k^{x_k}. \tag{10.7}$$

Expand the multinomial coefficient, multiply numerator and denominator by $(n - x_1)!$, and take $n!/x_1!$ outside the summation to get

$$n!/[x_1!(n - x_1)!]p_1^{x_1} \sum_{x_2 + x_3 + \cdots + x_k = n - x_1} (n - x_1)!/x_2!x_3! \ldots x_k! \, p_2^{x_2} p_3^{x_3} \cdots p_k^{x_k}. \tag{10.8}$$

The expression in the summation is $(p_2 + p_3 + \cdots + p_k)^{n-x_1} = (\sum_{i=2}^{k} p_i)^{n-x_1}$, which is $(1 - p_1)^{n-x_1}$. This gives the marginal PMF as

$$n!/[x_1!(n - x_1)!]p_1^{x_1}(1 - p_1)^{n-x_1} = \binom{n}{x_1} p_1^{x_1}(1 - p_1)^{n-x_1}, \tag{10.9}$$

which is easily seen to be $\text{BINO}(n, p_1)$. This means that the marginal distribution of x_i is $\text{BINO}(n, p_i)$ for $i \in \{1, 2, \ldots, n\}$. Proceeding similarly, it could be shown that the marginal distribution of any proper subset of 2 to $k - 1$ variables is also multinomial.

10.3.2 CONDITIONAL DISTRIBUTIONS

Conditional distributions of multinomials are more important as these are used in Expectation Maximization Algorithms (EMA), genome analysis, etc. They are obtained as follows:

$$f(x_1|X_2 = x_2, X_3 = x_3, \ldots) = f(x_1, x_2, \ldots, x_n)/f(x_2, x_3, \ldots, x_n), \tag{10.10}$$
$$f(x_1, x_2|X_3 = x_3, X_4 = x_4, \ldots) = f(x_1, x_2, \ldots, x_n)/f(x_3, x_4, \ldots, x_n), \tag{10.11}$$

which considerably simplifies when the variates are independent. For instance, the conditional distribution of X_j given that all other variables are held constant as $X_{i,i \neq j} = n_i$ is $\text{BINO}(N, p')$ where $N = n - \sum_{i \neq j = 1}^{k} n_i$ and $p' = p_j/(1 - \sum_{i \neq j = 1}^{k} p_i)$. Conditional joint distribution of r variables by keeping the rest $k - r$ variables constant is also multinomial $\text{MN}(n, p_1/P_r, p_2/P_r, \ldots, p_r/P_r)$ where P_r is the sum of the probabilities of $k - r$ variables held constant subtracted from one (which is the same as the sum of the probabilities of r variables).

Let X_n be a multinomial distribution with k classes defined above. Suppose we have missing data in an experiment. For convenience we assume that the first j components are observed, and $j + 1$ through k classes have missing data (unobserved). To derive the EMA for this type of problems, one needs to find the conditional distribution of $X \mid$ observed variates. The conditional distribution of X_i given $X_j = n_j$ is binomial with parameters $n-n_j$ and probability $p_i/(1 - p_j)$.

As $X_{j+1}, X_{j+2}, \ldots, X_k$ are unobserved with respective probabilities $p_{j+1}, p_{j+2}, \ldots, p_k$, we write it using $P(A|B) = P(A \cap B)/P(B)$ as

$$P\left[X_{j+1} = m_{j+1}, \ldots, X_k = m_k | X_1 = m_1, \ldots, X_j = m_j\right]$$
$$= \frac{P[X_1 = m_1, \ldots, X_k = m_k]}{P[X_1 = m_1, \ldots, X_j = m_j]}. \tag{10.12}$$

Due to the independence of the trials, this becomes

$$\frac{n!}{x_1! x_2! \ldots x_k!} \prod_{i=1}^{k} p_i^{x_i} / \frac{n!}{x_1! x_2! \ldots x_j!} \prod_{i=1}^{j} p_i^{x_i}. \tag{10.13}$$

Cancel out common terms to get a multinomial distribution.

10.3.3 GENERATING FUNCTIONS

The PGF for $t = (t_1, t_2, \ldots, t_k)$ is

$$P_x(t) = E(t^x) = E\left(\prod_{i=1}^{n} t_i^{x_i}\right) = \sum_{x=0}^{n} t^x \binom{n}{x_1, x_2, \ldots, x_k} p_1^{x_1} p_2^{x_2} \cdots p_k^{x_k}. \tag{10.14}$$

Write t^x as $t^{x_1+x_2+\cdots+x_k} = t_1^{x_1} t_2^{x_2} \ldots t_k^{x_k}$, then combine matching terms to get $(p_1 t_1 + p_2 t_2 + \cdots + p_k t_k)^n = (\sum_{j=1}^{k} p_j t_j)^n$.

PGF for any subset of variables could be obtained from the above by putting the corresponding dummy variables as 1 for the rest of them. Thus, the PGF of x_1, x_2 is $(p_1 t_1 + p_2 t_2 + \sum_{j=3}^{k} p_j)^n$. As the sum of the probabilities is one, this becomes $(p_1 t_1 + p_2 t_2 + 1 - (p_1 + p_2))^n = [1 + p_1(1 - t_1) + p_2(1 - t_2)]^n$. The PGF of a sum can be obtained by setting $t_i = t$ for all variables included in the summation. The MGF is obtained from this as $M_x(t) = (\sum_{j=1}^{k} p_j \exp(t_j))^n$, and the FMGF is obtained by replacing t by $(1 + t)$ as $(\sum_{j=1}^{k} p_j(1 + t_j))^n$. Raw moments (about 0) can also be obtained from the MGF. Thus, $E(X_j) = \partial M_x(t)/\partial t_j|_{t=0} = np_j(\sum_{j=1}^{k} p_j \exp(t_j))^{n-1}|_{t=0} = np_j$. The ChF is given by $\phi(t) = [\sum_{j=1}^{m} p_j(e^{it_j})]^n$. FCGF follows from it by taking log as $n \log(\sum_{j=1}^{k} p_j(1 + t_j))$.

Problem 10.2 If n is large and p_i is small such that $np_i = \lambda_i$ is constant, the multinomial distribution approaches $e^{-(\lambda_1+\lambda_2+\cdots+\lambda_k)} \lambda_1^{n_1} \lambda_2^{n_2} \cdots \lambda_k^{n_k} / [n_1! n_2! \cdots n_k!]$. *Hint:* The easiest way to prove this result is using PGF.

10.3.4   MOMENTS AND CUMULANTS

For each class the means can be obtained using binomial distribution as $E(X_i) = np_i$. The mean is directly obtained as

$$E(x_j) = \sum_{j=0}^{n} x_j \binom{n}{x_1, x_2, \ldots, x_k} p_1^{x_1} p_2^{x_2} \cdots p_k^{x_k}. \qquad (10.15)$$

As done in the case of binomial, expand the multinomial coefficient, take out n and p_j outside the summation to get

$$E(x_j) = np_j \sum_{j=0}^{n} \binom{n-1}{x_1, x_2, \ldots, x_k} p_1^{x_1} p_2^{x_2} \cdots p_j^{x_j-1} \cdots p_k^{x_k}. \qquad (10.16)$$

The summation evaluates to $(p_1 + p_2 + \cdots + p_k)^{n-1} = 1$, giving $E(x_j) = np_j$. To evaluate the product moment $E(X_i X_j)$, write $X_i X_j = [(X_i + X_j)^2 - X_i^2 - X_j^2]/2$, and take expectation of both sides to get $E(X_i X_j) = n(n-1)p_i p_j$, from which the variance is $np_j(1 - p_j)$, and the covariance is $-np_i p_j$ for $i \neq j$. As the covariance is negative, so is the correlation. This is because when one of the variables increases, the other must decrease due to the sum constraint on the $X_i's$. Due to the presence of factorial terms in the denominator of the multinomial coefficient, factorial moments are easier to evaluate. The combined factorial moment is

$$\mu'_{r_1, r_2, \ldots, r_k} = \sum x1_{(r_1)} x2_{(r_2)} \ldots xk_{(r_k)} \binom{n}{x_1, x_2, \ldots, x_k} \prod_{i=1}^{k} p_i^{n_i}, \qquad (10.17)$$

where $xi_{(r_i)}$ denotes the Pochhammer falling factorial. Expand the multinomial coefficient and cancel out common factors to get $\mu'_{r_1, r_2, \ldots, r_k} = n_{(\sum_{j=1}^{k} r_j)} \prod_{j=1}^{k} p_j^{r_j}$. As in the case of binomial, $Z_j = (X_j - np_j)/[p_j(1 - p_j)]^{1/2}$ asymptotically approaches the standard normal distribution $N(0, 1)$, so that $Z = (Z_1, Z_2, \ldots, Z_k)$ approaches multivariate normal distribution. Cumulants are easier to find using the CGF $K_x(t) = \log(M_x(t)) = n \log(\sum_{j=1}^{k} p_j \exp(t_j))$.

Example 10.3   Relation to Poisson distribution   If $X_j \sim$ IID POIS(λ) for $j = 1, 2, \ldots, n$; and $S = X_1 + X_2 + \cdots + X_n$ then $\Pr[X|S]$ has a multinomial distribution where $X = (X_1, X_2, \ldots, X_n)$.

Solution: Because of independence (Chapter 6) $S \sim$ POIS($n\lambda$) with PMF $\exp(-n\lambda)(n\lambda)^s/s!$. The joint distribution of X is $\prod_{j=1}^{n} \exp(-\lambda)\lambda^{x_j}/x_j! = \exp(-n\lambda)\lambda^{\sum x_j}/\prod_{j=1}^{n} x_j!$. From this the conditional distribution is obtained as

$(\sum_{j=1}^{n} x_j)! / \prod_{j=1}^{n} x_j! \times (1/n)^{\sum_{j=1}^{n} x_j} = s! / [n^s \prod_{j=1}^{n} x_j!]$, which is a multinomial distribution with each probability $p = 1/n$ and $n = \sum_{j=1}^{n} x_j$. In general, if $X_j \sim$ IID POIS(λ_j), then $\Pr[X|S] \sim$ MN($n, p_1', p_2', \cdots, p_k'$) where $p_j' = p_j / \sum_{i=1}^{k} p_i$.

Example 10.4 Poisson decomposition Suppose the total number of events $X \sim$ POIS(λ), where each of the events can be classified as mutually independent types X_1 of Type-1 with probability p_1, X_2 of Type-2 with probability p_2, and so on, X_m of Type-m with probability p_m. Then $X_k \sim$ IID POIS(λp_k) for $k = 1, 2, \ldots, m$.

Solution: Let $n_1, n_2, \ldots, n_m$ be the number of events of each type. Then $n_1 + n_2 + \cdots + n_m = n$ is the total number of events. Joint distribution of X_k conditional on $N = n$ is multinomial with PMF

$$\Pr(X_1 = x_1, X_2 = x_2, \ldots, X_m = x_m | X = n) = \binom{n}{n_1, n_2, \ldots, n_m} p_1^{x_1} p_2^{x_2} \cdots p_m^{x_m} \quad (10.18)$$

so that $p_1 + p_2 + \cdots + p_m = 1$. As $X \sim$ POIS(λ), $\Pr[X = n] = \exp(-\lambda)\lambda^n / n!$. The unconditional distribution of $X_1, X_2, \ldots, X_m$ is obtained by multiplying as

$$\Pr(X_1 = x_1, X_2 = x_2, \ldots, X_m = x_m) = \binom{n}{n_1, n_2, \ldots, n_m} p_1^{x_1} p_2^{x_2} \cdots p_m^{x_m} \exp(-\lambda)\lambda^n / n!. \quad (10.19)$$

Cancel out $n!$ from numerator and denominator, write n as $n_1 + n_2 + \cdots + n_m$, and $\exp(-\lambda)$ as $\exp(-\lambda(p_1 + p_2 + \cdots + p_m)) = \prod_{k=1}^{m} \exp(-\lambda p_k)$. Combine like terms to get

$$\Pr(X_1 = x_1, X_2 = x_2, \ldots, X_m = x_m)$$
$$= \exp(-\lambda p_1) \exp(-\lambda p_2) \ldots \exp(-\lambda p_k)(\lambda p_1)^{x_1} (\lambda p_2)^{x_2} \ldots (\lambda p_m)^{x_m} / [x_1! x_2! \ldots x_m!]. \quad (10.20)$$

As this can be written as $\prod_{k=1}^{m} \exp(-\lambda p_k)(\lambda p_k)^{x_k} / x_k!$, each of the $X_k \sim$ POIS(λp_k).

Example 10.5 Human blood groups Suppose the percentage of people with the blood groups {A, B, O, AB} are 40, 12, 5, and 43, respectively. Find the probability that (i) in a group of 60 students, 30 or more are of blood group "A", and (ii) at least 4 persons have blood group O.

Solution: Using the frequency approach, we expect the probability of any person with blood group "A" as 0.40. Denote this as $p_1 = 0.40$. Similarly, $p_2 = 0.12$, $p_3 = 0.05$, and $p_4 = 0.43$. This gives the PMF as

$$f(x) = 60! / [x_1! x_2! x_3! x_4!](.40)^{x_1} (.12)^{x_2} (.05)^{x_3} (.43)^{x_4} \quad (10.21)$$

such that $x_1 + x_2 + x_3 + x_4 = 60$. As the marginal distribution is binomial, the answer to (i) is $\sum_{x=30}^{60} \text{BINO}(60, 0.40) = 0.074624$, and the (ii) probability of "O" blood group is $5/100 = 1/20$. Thus, the answer is $1 - \sum_{i=0}^{3} \text{BINO}(60, 1/20) = 1 - \sum_{i=0}^{3} \binom{60}{i}(1/20)^i (19/20)^{60-i} = 1 - 0.647281 = 0.352719$.

10.4 APPLICATIONS

As it is an extension of the binomial distribution to more than two classes, it finds lots of applications where the number of mutually exclusive and independent classes are three or more. Fisher *et. al.* (1943)[25] used the univariate LSD to model species abundance. Due to inter-dependency among multiple species in a bounded region, there is an increasing interest for multiple species modeling using multinomial distribution due to the possibility of comparing observed covariance in species abundance to the expected. Sequences of multinomial random vectors are used in nonparametric statistics, estimation, queuing theory, information retrieval, data science, etc. It is used in computational biology (biological sequence analysis, plant breeding using fixed phenotypes), bioinformatics (protein synthesis and structure comparisons), data communications (transmission error types), pattern recognition (matching patterns into known groups or categories), marine biology (parasite infestation of creatures like crabs, shrimps), particle physics and statistical thermodynamics (micro-states of particles), to name a few. An application to corona screening at entry points appears in Shanmugam (2020) [91], where three groups called *corona ill, healthy from corona*, and *pending* are considered for travelers subjected to diagnostic tests.

10.4.1 CONTINGENCY TABLES

The joint distribution of two or more categorical variables, each at different levels is usually captured into a table called contingency table. They are identified using the $r \times c$ format where r is the number of rows (categories or groups), and c is the number of columns (characteristics or features). Thus, 2×2 denotes a contingency table with 2 rows and 2 columns. Under the null hypothesis that characteristics are independent of categories, the $\chi^2_{(r-1)(c-1)}$ test can be used for inference [67]. The χ^2 test is applicable when the frequencies are assumed to follow a multinomial distribution. Thus the joint distribution of counts n_{ij} observed in the ith row and jth column, as well as the conditional distribution of counts conditional on the values of $n_{i.}$ (sum of the counts on ith row) and $n_{.j}$ (sum of the counts on jth column) are multinomial distributed.

10.4.2 GENOMICS

This distribution is used for various purposes in genomics (genotype matching, DNA sequencing, allele frequencies of genes into allelic types, etc.). Consider two types of genes (denote them by a and A) present at a fixed locus in a chromosome pair. Then possible genotype combinations are {aa, aA, Aa, AA}. As aA and Aa are identical, we have only three categories {aa, aA, AA}. Let p denote the proportion of a, and $q = 1 - p$ denote the proportion of A. Due to the independence of chromosome pair, we could assign probabilities as $\Pr[\text{aa}] = \pi_1 = p^2, \Pr[\text{AA}] = \pi_2 = q^2$, and $\Pr[\text{aA}] = \pi_3 = 2pq$. These are the probabilities of selecting two alleles at random from the pool. Assume that these serve as bins. Now select a sample of n individuals from the population and classify them into the bins. Let X_k denote the number of individuals falling in each bin. Then $X_1, X_2, X_3 \sim \text{MN}(n, \pi_1, \pi_2, \pi_3)$. Although there are three unknown probabilities (π_1, π_2, π_3), all of them are functions of p (as $q = 1 - p$). Such models are called reduced-parameter models, which underlie the Hardy–Weinberg equilibrium.

10.4.3   FINANCE

Financial analysts use the multinomial distribution for a variety of purposes. Examples are in asset allocations, investment risk modeling, identifying types of loan defaulters, predicting exchange rate fluctuations (for selected currencies), risk analysis of stocks in portfolios, performance of a given set of portfolio returns, etc. These categories or classes are considered to be independent, the probabilities of which are estimated from past data. Building a multinomial model is then straightforward.

10.4.4   RELIABILITY

Complex hardware and software systems comprising of multiple independent components may fail due to a variety of reasons. These components may come from different manufacturers in the case of hardware systems. If the probabilities of failure of these components are known from past data, we could evaluate any particular combination of failures using the multinomial distribution [6]. In the case of software systems, these components might have been developed by different groups, or using various technologies like multiple programming languages or paradigms. If the probability of events of interest are known in advance, a multinomial model can be developed.

10.5   SUMMARY

The multinomial distribution is as popular in discrete distributions as multivariate normal distribution is in continuous case. Basic properties of multinomial distribution are described in this chapter. Some applications in different fields are also mentioned.

Bibliography

[1] Allen, A. O. (2014). *Probability, Statistics and Queuing Theory with Computer Science Applications*, Academic Press, Elsevier. DOI: 10.2307/1269262. 106

[2] Anscombe, F. J. (1950). Sampling theory of the negative binomial and logarithmic series distributions, *Biometrika*, 37:358–382. DOI: 10.1093/biomet/37.3-4.358. 121, 170

[3] Arratia, R., Goldstein, L., and Gordon, L. (1989). Two moments suffice for Poisson approximation, *The Annals of Probability*, 17(1):9–25. http://bcf.usc.edu/~larry/papers/pdf/AGG.pdf 108

[4] Avlo, M. and Cabilio, P. (2000). Calculation of hypergeometric probabilities using Chebyshev polynomials, *The American Statistician*, 54(2):141–144. DOI: 10.1080/00031305.2000.10474527. 139

[5] Balakrishnan, N. and Nevzorov, V. B. (2003). *A Primer on Statistical Distributions*, John Wiley, NY. DOI: 10.1002/0471722227. 1, 151

[6] Balakrishnan, N. and Rao, C. R. (2001). *Handbook of Statistics*, vol. 20, Advances in Reliability, Elsevier. DOI: 10.1198/tech.2003.s180. 188

[7] Bardsley, W. E. (2016). Note on the hypergeometric distribution as an invalidation test for binary forecasts, *Stochastic Environmental Research and Risk Assessment*, 30(3):1059–1061. DOI: 10.1007/s00477-015-1071-z. 145

[8] Bartolucci, A. A., Shanmugam, R., and Singh, J. (2001). Development of the generalized geometric model with application to cardiovascular studies, *Systems Analysis Modelling and Simulation*, 41:339–349. https://www.researchgate.net 78

[9] Bonetti, M., Cirillo, P., and Ogay, A. (2019). Computing the exact distributions of some functions of the ordered multinomial counts: Maximum, minimum, range and sum of order statistics, *Journal of Royal Society*. DOI: 10.1098/rsos.190198. 181

[10] Boswell, M. T. and Patil, G. P. (1970). Chance mechanisms generating the negative binomial distributions, *Random Counts in Scientific Work*, 1:3–22. Pennsylvania State University. 86, 99

[11] Bravo de Guenni, L. (2011). Poisson distribution and its application in statistics, *International Encyclopedia of Statistical Science*. DOI: 10.1007/978-3-642-04898-2_448. 126

[12] Carr,W.E. (1980). Fisher's exact test extended to more than two samples of equal size, *Technometrics*, 22:269–270. 146

[13] Chapman, D. G. (1951). Some properties of the hypergeometric distribution with applications to zoological sample censuses, *University of California Publication in Statistics*, 1:131–159. 148

[14] Chattamvelli, R. (1995). A note on the noncentral beta distribution function, *The American Statistician*, 49:231–234. DOI: 10.2307/2684647. 120

[15] Chattamvelli, R. (2012). *Statistical Algorithms*, Alpha Science, Oxford, UK. 43, 77

[16] Chattamvelli, R. (2016). *Data Mining Methods*, 2nd ed., Alpha Science, Oxford, UK. 150

[17] Chattamvelli, R. and Jones, M. C. (1996). Recurrence relations for noncentral density, distribution functions, and inverse moments, *Journal of Statistical Computation and Simulation*, 52(3):289–299. DOI: 10.1080/00949659508811679. 27

[18] Chattamvelli, R. and Shanmugam, R. (2019). *Generating Functions in Engineering and the Applied Sciences*, Morgan & Claypool. DOI: 10.2200/s00942ed1v01y201907eng037. 12, 23, 37, 75, 89, 90, 91, 143, 164

[19] Consul, P. C. (1974). A simple urn model dependent upon predetermined strategy, *Sankhya, B*, 36:391–399. 45

[20] Consul, P. C. (1989). *Generalized Poisson Distributions*, Marcel Dekker, NY. 121

[21] Consul, P. C. (1990). On some properties and applications of quasi-binomial distributions, *Communications in Statistics-Theory and Methods*, 19:477–504. DOI: 10.1080/03610929008830214. 45

[22] Consul, P. C. and Shenton, L. R. (1972). Use of Lagrange expansion for generating generalized probability distributions, *SIAM Journal of Applied Mathematics*, 23:239–248. DOI: 10.1137/0123026. 45, 99

[23] Cournot, M. A. (1843). Exposition de la theorie des chances et des probabilites, *Librairie de L. Machette*, Paris. 135

[24] Fahidy, T. Z. (2012). Application of the negative binomial/Pascal distribution in probability theory to electrochemical processes, *Recent Trends in Electrochemical Science and Technology*, Intech Open. DOI: 10.5772/33946. 99

[25] Fisher, R. A., Corbet, A. S., and Williams, C. B. (1943). The relation between the number of species and the number of individuals in a random sample of an animal population, *Journal of Animal Ecology*, 12(1):42–58. DOI: 10.2307/1411. 177, 187

[26] Forbes, C., Evans, M., Hasting, N., and Peacock, B. (2011). *Statistical Distributions*, 4th ed., Wiley, NY. DOI: 10.1002/9780470627242. 1

[27] Godbole, A. P. (1990). On hypergeometric and related distributions of order k, *Communications in Statistics–Theory and Methods*, 19(4):1291–1301. DOI: 10.1080/03610929008830262. 140, 164

[28] Guenther, W. C. (1975). The inverse hypergeometric distribution—a useful model, *Statistica Neerlandica*, 29:129–144. DOI: 10.1111/j.1467-9574.1975.tb00257.x. 159

[29] Guenther, W. C. (1978). Some remarks on the runs test and the use of the hypergeometric distribution, *The American Statistician*, 32(2):71–73. DOI: 10.2307/2683620. 139, 146, 159

[30] Gurmu, S. and Trivedi, P. K. (1992). Overdispersion tests for truncated Poisson regression models, *Journal of Econometrics*, 54:347–370. DOI: 10.1016/0304-4076(92)90113-6. 98

[31] Hidiroglou, M. A. (1978). An approx of the inverse moments of the positive hypergeometric distribution, *Communications in Statistics–Theory and Methods*, 7(15):1475–1487. DOI: 10.1080/03610927808827729. 142

[32] Hilbe, J. (2011). *Negative Binomial Regression*, Cambridge University Press, NY. DOI: 10.1017/cbo9780511973420.009. 99

[33] Hu, D. P., Cui, Y. Q., and Yin, A. H. (2013). An improved negative binomial approximation for negative hypergeometric distribution, *Applied Mechanics and Materials*, 427–429. DOI: 10.4028/www.scientific.net/amm.427-429.2549. 158

[34] Irwin, J. O. (1954). A distribution arising in the study of infectious diseases, *Biometrika*, 41:266–268. DOI: 10.1093/biomet/41.1-2.266. 158

[35] Jain, G. C. and Consul, P. C. (1971). A generalized negative binomial distribution, *SIAM Journal of Applied Mathematics*, 21:501–503. DOI: 10.1137/0121056. 45, 99

[36] Jain, G. C. and Gupta, R. P. (1973). A logarithmic series distribution, *Trab. Estadistica*, 24:99–105. DOI: 10.1007/bf03013757. 170

[37] Johnson, N. L. (1957). A note on the mean deviation of the binomial distributions, *Biometrika*, 44:532–533. DOI: 10.1093/biomet/45.3-4.587. 13, 33

[38] Johnson, N. L., Kotz, S., and Balakrishnan, N. (2004). *Discrete Multivariate Distributions*, 2nd ed., John Wiley, New York. 181

[39] Johnson, N. L., Kemp, A. W., and Kotz, S. (2005). *Univariate Discrete Distributions*, 3rd ed., John Wiley, NY. DOI: 10.1002/0471715816. 1, 121, 151

[40] Jones, S.N. (2013). A gaming application of the negative hypergeometric distribution, UNLV graduate thesis, Dept. of Math. sciences. https://digitalscholarship.unlv.edu/thesesdissertations/1846 158

[41] Kamat, A. R. (1966). A generalization of Johnson's property of the mean deviation for a class of discrete distributions, *Biometrika*, 53:285–287. DOI: 10.2307/2334086. 13, 144, 164

[42] Kastner, M., et al. (2009). The capture-mark-recapture technique can be used as a stopping rule when searching in systematic reviews, *Journal of Clinical Epidemiology*, 62(2):149–157. DOI: 10.1016/j.jclinepi.2008.06.001. 150

[43] Kemp, C. D. and Kemp, A. W. (1956). Generalized hypergeometric distributions, *Journal of Royal Statistical Society, Series B*, 18:202–211. DOI: 10.1111/j.2517-6161.1956.tb00224.x. 158

[44] Khan, R. A. (1994). A note on the generating function of a negative hypergeometric distribution, *Sankhya, B*, 56:309–313. 159

[45] Khang, T. F. and Ong, S. H. (2007). A new generalization of the logarithmic distribution arising from the inverse trinomial distribution, *Communications in Statistics–Theory and Methods*, 36(11):3–21. DOI: 10.1080/03610920600966480. 172

[46] King, R. and McCrea, R. (2019). Capture-recapture methods and models: Estimating population size, *Handbook of Statistics*, 40(2):33–83. DOI: 10.1016/bs.host.2018.09.006. 145

[47] Knoll, G. F. (2017). *Radiation Detection and Measurement*, 4th edition, John Wiley, NY. 116

[48] Knuth, D. E. (2011). *Fundamental Algorithms*, vol. 1, Addison Wesley https://www-cs-faculty.stanford.edu/~knuth/taocp.html. 43, 110

[49] LeCam, L. (1960). An approximation theorem for the Poisson binomial distribution, *Pacific Journal of Mathematics*, 10:1181–1197. Thermal Nanosystems and Nanomaterials, Springer. DOI: 10.2140/pjm.1960.10.1181.

[50] Lee, P.A. and Ong, S.H. (1986). The bivariate noncentral negative binomial distributions, *Metrika*, 33:1–28. 6

[51] Ling, K. D. (1993). Sooner and later waiting distributions for frequency quota defined on a Polya-Eggenberger urn model, *Soochow Journal of Mathematics*, 19(2):139–151. 158

[52] Lopez-Blazquez, F. and Salamanca Mino, B. (2000). Binomial approximation to hypergeometric probabilities, *Journal of Statistical Planning and Inference*, 87:21–29. DOI: 10.1016/s0378-3758(99)00187-1. 140

[53] Lord, F. M. and Novick, M. R. (1962). *Statistical Theories of Mental Test Scores*, Information Age Publishing. 158

[54] Mathai, A. M. and Haubold, H. J. (2008). *Special Functions for Applied Scientists*, Springer. DOI: 10.1007/978-0-387-75894-7. 137, 138

[55] Mehta, C. R. and Patel, N. R. (1983). A network algorithm for performing Fisher's exact test in RxC contingency tables, *Journal of the American Statistical Association*, 78:427–434. DOI: 10.2307/2288652. 146

[56] Miller, G.K., and Fridell, S.L. (2007) A Forgotten Discrete Distribution? Reviving the Negative Hypergeometric Model *The American Statistician*, Vol. 61(4), 347-350, https://www.jstor.org/stable/27643937.

[57] Molenaar, W. (1973). Simple approximations to the Poisson, binomial and hypergeometric distributions, *Biometrics*, 29:403–408. DOI: 10.2307/2529405. 140

[58] Ong, S. H. and Lee, C. M. S. (1979). *The Noncentral Negative Binomial Distribution*, Biometrical journal, 21(7), 611-627. DOI: 10.1002/bimj.4710210704

[59] Ong, S. H. (1987). Some notes on the non-central negative binomial distribution, *Metrika*, 34:225–236. DOI: 10.1007/bf02613154. 6

[60] Ord, J. K. (1967). Graphical methods for a class of discrete distributions, *Journal of Royal Statistical Society, Series A*, 130:232–238. DOI: 10.2307/2343403. 170

[61] Papoulis, A. and Unnikrishna Pillai, S. (2017). *Probability, Random Variables and Stochastic Processes*, McGraw Hill.

[62] Patnaik, P. B. (1949). The noncentral chi-square and F-distributions and their applications, *Biometrika*, 46:202–232. DOI: 10.1093/biomet/36.1-2.202. 108

[63] Phillips, T. R. L. and Zhigljavsky, A. (2014). Approximation of inverse moments of discrete distributions, *Statistics and Probability Letters*, 94:135–143. DOI: 10.1016/j.spl.2014.07.007. 142

[64] Plachky, D. (2003). Relationships between the negative binomial and negative hypergeometric distributions with applications to testing and estimation, *American Journal of Mathematical and Management Sciences*, 23(1):1–6. DOI: 10.1080/01966324.2003.10737601. 158

[65] Quenouille, M. H. (1949). A relation between the logarithmic, Poisson and negative binomial series, *Biometrics*, 5(2):162–164. DOI: 10.2307/3001917. 171

[66] Rao, C. R. (1971). Some comments on the logarithmic series distribution in the analysis of insect trap data, *Statistical Ecology*, 1:131–142, PSU Press, Pennsylvania. 177

[67] Rao, C. R. (1973). *Linear Statistical Inference and its Applications*, 2nd ed., John Wiley, NY. DOI: 10.1002/9780470316436. 121, 132, 181, 187

[68] Robson, D. S. and Regier, H. A. (1964). Estimation of population number and mortality rates, *IBP Handbook*, no. 3, Blackwell, Oxford. 148

[69] Roy, S., Tripathi, R. C., and Balakrishnan, N. (2019). A Conway Maxwell Poisson type generalization of the negative hypergeometric distribution, *Communications in Statistics–Theory and Methods*. DOI: 10.1080/03610926.2019.1576885. 164

[70] Sadinle, M. (2008). Linking the negative binomial and logarithmic series distributions via their associated series, *Revista Columbiana de Estadistica*, 31(2):311–319. https://www.researchgate.net 171

[71] Sadygov, R.G. and Yates, J.R. (2003). A hypergeometric probability model for protein identification and validation using tandem mass spectral data and protein sequence databases, *Analytical Chemistry*, 75(15):3792–3798. DOI: 10.1021/ac034157w. 151

[72] Schuster, E. F. and Sype, W. R. (1987). On the negative hypergeometric distribution, *International Journal of Mathematics Education in Science and Technology*, 18(3):453–459. DOI: 10.1080/0020739870180316. 161

[73] Seber, G. A. F. (2002). *The Estimation of Animal Abundance and related parameters*, 2nd edition, The Blackburn Press. 148

[74] Shanmugam, R. (1982). A characterization of negative binomial distribution truncated at zero, *Journal of the Korean Statistical Society*, 11(2):131–138. DOI: 10.1080/02331888608801906. 98

[75] Shanmugam, R. (1985). An intervened Poisson distribution and its medical application, *Biometrics*, 41:1025–1029. DOI: 10.2307/2530973. 121

[76] Shanmugam, R. (1988). A goodness of fit test for the Lagrangian logarithmic series distribution, *Journal of Statistics Research*, 22(1,2):31–36. 176

[77] Shanmugam, R. (1991). Incidence rate restricted Poissonness, *Sankhya, B*, 53(2):191–201. https:/www.jstor.org/stable/25052691 121

[78] Shanmugam, R. (1992). An inferential procedure for the Poisson intervention parameter, *Biometrics*, 48:559–565. DOI: 10.2307/2532309. 122

[79] Shanmugam, R. (2001). Predicting a "successful" prevention of an epidemic, *Communications in Statistics–Theory and Methods*, 30(1):93–103. DOI: 10.1081/sta-100001561. 121, 125

[80] Shanmugam, R. (2006). Poisson distribution, *Encyclopedia of Measurement and Statistics*, vol. 2, N. J. Salkind, Ed., 772–776. DOI: 10.4135/9781412952644. 101, 121, 126

[81] Shanmugam, R. (2011). Spinned Poisson distribution with health management application, *Healthcare Management Science*, 14:299–306. DOI: 10.1007/s10729-011-9157-8. 122

[82] Shanmugam, R. (2012). Intervened: 2-tier Poisson distribution for understanding hospital site infectivity, *International Journal of Research in Nursing*, 3(1):8–14. DOI: 10.3844/ijrnsp.2012.8.14. 121

[83] Shanmugam, R. (2013). Odds to quicken reporting already delayed cases: AIDS incidences are illustrated, *International Journal of Research in Nursing*, 4(1):1–13. DOI: 10.3844/ijrnsp.2013.1.13. 78

[84] Shanmugam, R. (2013). Does smoking delay pregnancy? Data analysis by a tweaked geometric distribution answers, *International Journal of Research in Medical Sciences*, 1:343–348. DOI: 10.5455/2320-6012.ijrms20131106. 78

[85] Shanmugam, R. (2013). Does over or under dispersion in inverse binomial data suggest anything?, *American Journal of Biostatistics*, 3(2):30–37. DOI: 10.3844/amjbsp.2013.30.37. 99

[86] Shanmugam, R. (2013). Shortage level of matching kidney and pancreas organs for implant is estimated, *International Journal of Research in Nursing*, 4(2):40–46. DOI: 10.3844/ijrnsp.2013.40.46. 99

[87] Shanmugam, R. (2014). Paradox in earthquake free safety vs. "tectonic glue" based on geometric pattern of 21st century indices, *SSRG International Journal of Geoinformatics and Geological Sciences*, 10(1):1–4. internationaljournalssrg.org 102

[88] Shanmugam, R. (2015). Curvature informatics about medical errors: Geometric view of incidence restricted Poisson, *International Journal of Ecological Economics and Statistics*, 36(4):1–10. ceserp.com/cp-jour, ceserpublications.com 121

[89] Shanmugam, R. (2015). Never, once, and repeated illness: Geometric view for insights and interpretations, *International Journal of Research in Medical Sciences*, 3(6):1336–1341. DOI: 10.18203/2320-6012.ijrms20150142. 99

[90] Shanmugam, R. (2016). Could geometry demystify patterns of earthquakes and after-shocks?, *American Journal of Geo-Sciences*, 6(1):1–9. DOI: 10.3844/ajgsp.2016.1.9. 102

[91] Shanmugam, R. (2020). Distracted multinomial model for corona screening at entry ports, *International Journal of Research in Medical Sciences*, 1–8. DOI: 10.18203/2320-6012.ijrms20201904. 187

[92] Shanmugam, R., Johnson, C., and Cutter, G. (2006). Examining whether an epidemic is excessive, *Journal of Statistics and Applications*, 1(1):51–62. 122

[93] Shanmugam, R. and Radhakrishnan, R. (2011). Incidence jump rate reveals under/over dispersion in count data, *International Journal of Data Analysis and Information Systems*, 3(1):1–9. 27, 69, 99

[94] Shanmugam, R. and Singh, J. (1984). A characterization of the logarithmic series distribution and its application, *Communications in Statistics–Theory and Methods*, 13(7):865–875. DOI: 10.1080/03610928408828725. 171, 176

[95] Shanmugam, R. and Singh, J. (1995). A weighted logarithmic series model for purchasing behavior, *Journal of Applied Statistical Science*, 2(1):61–71. 171

[96] Shanmugam, R. and Singh, K. (2001). Testing of incidence rate restriction, *International Journal of Reliability and Applications*, 2(4):263–268. 121

[97] Shanmugam, R. and Singh, J. (2003). Index of variation with illustration using molecular data, *Communications in Statistics–Theory and Methods*, 32(2):509–517. DOI: 10.1081/sta-120018198. 21

[98] Shonkwiler, J. S. (2016). Variance of the truncated negative binomial distribution, *Journal of Econometrics*, 195(2):209–210. DOI: 10.1016/j.jeconom.2016.09.002. 98

[99] Song, W. T. (2005). Relationships among some univariate distributions, *IEEE Transactions*, 37:651–656. DOI: 10.1080/07408170590948512. 67, 86

[100] Student. (1907). On the probable error of the mean, *Biometrika*. DOI: 10.2307/2331554.

[101] Tate,R.F., and Goen,R.L.(1958). Minimum variance unbiased estimation for the truncated Poisson distribution, *Annals of Mathematical Statistics*, 29:755–765. 125

[102] Teerapabolarn, K. (2011). On the Poisson approximation to the negative hypergeometric distribution, *Bulletin of the Malaysian Mathematical Sciences Society*, 34(2):331–336. 158

[103] Teerapabolarn, K. (2015). Negative binomial approximation to the beta binomial distribution, *International Journal of Pure and Applied Mathematics*, 98(1):39–43. DOI: 10.12732/ijpam.v98i1.5. 158, 164

[104] Vellaisamy, P. and Punnen, A. P. (2001). On the nature of the binomial distribution, *Journal of Applied Probability*, 38(1):36–44. DOI: 10.1239/jap/996986641. 45

[105] Watterson, G. A. (1974). Models for the logarithmic species abundance distributions, *Theoretical Population Biology*, 6(2):217–250. DOI: 10.1016/0040-5809(74)90025-2. 177

[106] Wright, T. (2012). *Exact Confidence Bounds When Sampling from Small Finite Universes: An Easy Reference Based on the Hypergeometric Distribution*, Springer. DOI: 10.1007/978-1-4612-3140-0.

[107] Zhao, J., Zhang, Z., and Sullivan, C. J. (2019). Identifying anomalous nuclear radioactive sources using Poisson kriging and mobile sensor networks, *PLoS One*, 14(5). DOI: 10.1371/journal.pone.0216131. 116

Authors' Biographies

RAJAN CHATTAMVELLI

Rajan Chattamvelli Rajan Chattamvelli is a professor in the school of advanced sciences at VIT University, Vellore, Tamil Nadu. He has published more than 22 research articles in international journals of repute and at various conferences. His research interests are in computational statistics, design of algorithms, parallel computing, object-oriented programming, data mining, machine learning, combinatorics, and big data analytics. His prior assignments include Denver Public Health, Colorado; Metromail Corporation, Lincoln, Nebraska; Frederick University, Cyprus; Indian Institute of Management; Periyar Maniammai University, Thanjavur; and Presidency University, Bangalore.

RAMALINGAM SHANMUGAM

Ramalingam Shanmugam is a honorary professor in the school of Health Administration at Texas State University. He is the editor of the journals *Advances in Life Sciences, Global Journal of Research and Review, International journal of research in Medical Sciences,* and book-review editor of the *Journal of Statistical Computation and Simulation.* He has published more than 200 research articles and 120 conference papers. His areas of research include theoretical and computational statistics, number theory, operations research, biostatistics, decision making, and epidemiology. His prior assignments include University of South Alabama, University of Colorado at Denver, Argonne National Labs, Indian Statistical Institute, and Mississippi State University. He is a fellow of the International Statistical Institute.

Index

Printed in the United States
by Baker & Taylor Publisher Services